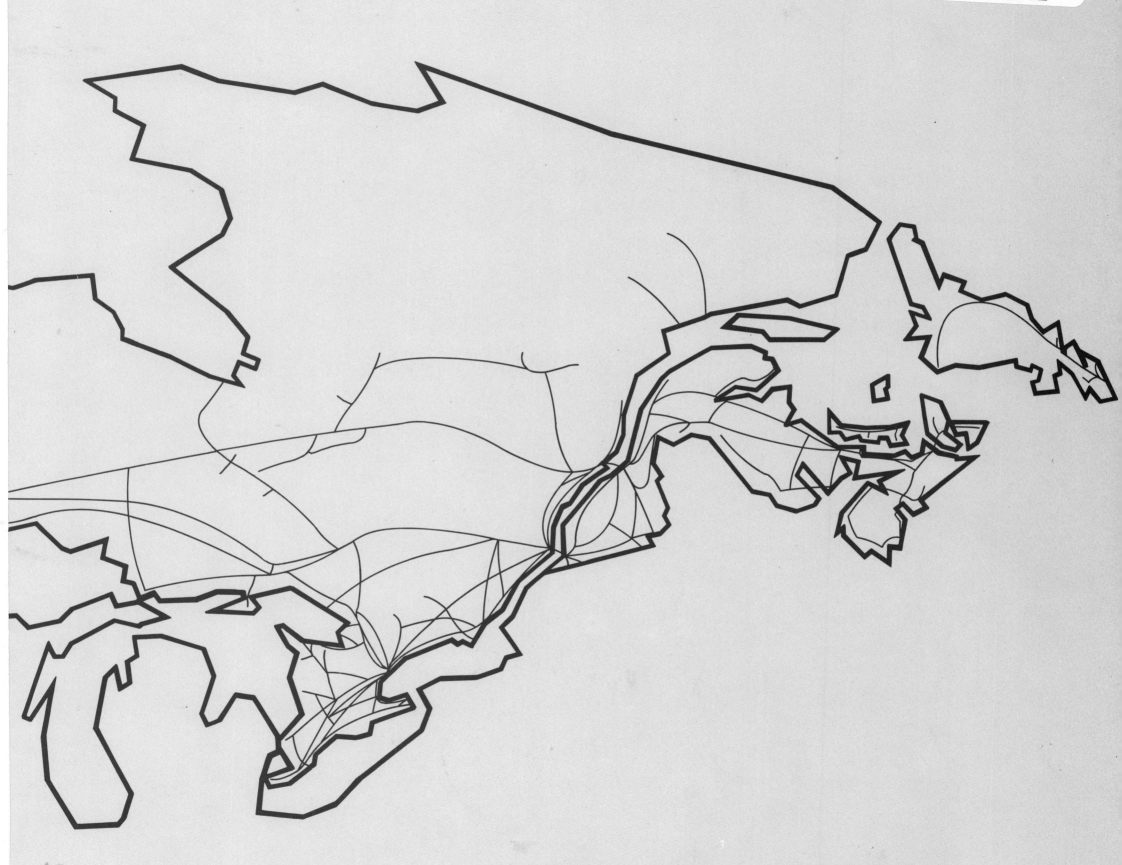

For Pauline Smaill, and all the dreams confided in the kitchen at 3495.

Signatures IN STEEL

Greg McDonnell

A Stoddart/Boston Mills Press Edition

Canadian Cataloguing in Publication Data

McDonnell, Greg, 1954 –
Signatures in Steel
"A Boston Mills Press edition".

ISBN 0-7737-2554-7

1. Railroads – Canada – Pictorial works.
I. Title.

TF26.M33 1991   385'.0971   C91-094195-5

Book Design and Production,
George W. Roth
Polygon Design Limited
Copy Editing,
Noel Hudson
Typesetting,
Dobbie Graphics Inc.
Printing and Binding,
Khai Wah Litho, Singapore

A Boston Mills Press Edition
First published in 1991 by
Stoddart Publishing Co. Limited
34 Lesmill Road
Toronto, Canada
M3B 2T6

Jacket photographs:
Front:
Flying green and running as First 904, CP SD40 5409
"The Whale" screams through Orr's Lake, Ontario, at
16:30 on April 21, 1986. Engineer G. Dagelman has the
ex-QNS&L SD in the company notch, and 24 cars back,
the photographer's Oakville-bound Ford pick-up truck is
safely secured aboard TTBX 912861.
Greg McDonnell

Back:
TH&B 82 brings up the markers as CP Rail's London-
Hamilton "Aberdeen Turn" drifts down Orr's Lake hill
west of Galt, Ontario, on June 13, 1990.
Greg McDonnell

Page 1
GO Transit bi-levels at Willowbrook Yard, Etobicoke,
Ontario, September 24, 1986.
Greg McDonnell

Page 2
CP 4217, 4738, 5537 crossing Grand River bridge with
#921 at Galt, Ontario, 07:15 September 20, 1990.
Greg McDonnell

Page 3
CN Work Extra 4528 Snowplow near Hensall, Ontario,
16:12 December 11, 1977.
Greg McDonnell

Where's 40? . . .

On a cool spring evening more than 30 years ago, my brother and I sat on an old wooden baggage cart at the CNR Kitchener, Ontario station, impatiently swinging our legs and asking, "Where's 40?" Due into town at 9:20 p.m., "40" was the nightly Sarnia-Toronto passenger train, just an unnamed local in the timecard, but for two young boys waiting on the platform at Kitchener, a nightly event of great importance.

The waiting room was crowded with passengers and the hum of a dozen conversations was punctuated by the resounding bang of the validator every time the ticket agent issued passage to Guelph, West Toronto, Union, Port Hope, Belleville, Kingston, Montreal, or countless faraway destinations. Beyond the wickets, the operator scribbled out train orders and rattled off repeats in an almost unintelligible but rhythmic chant as an ancient Seth Thomas clock – still neatly lettered G.T.R. in gold leaf – ticked off the minutes until train time. In the baggage room, Dad visited with the baggageman, who busily tagged luggage and hoisted it aboard the cart. They talked of diesels and politics and layoffs, while we pestered to be weighed on the big scale.

Outside, the yard engine, an 0-6-0 of Grand Trunk lineage, shunted cars at the freight sheds and then positioned itself to add a car of express to #40 upon arrival. Meanwhile, a bright red Royal Mail truck pulled up onto the platform and the suspense grew. Where was 40? There was a sense of urgency and an air of importance about all this activity, and we felt privileged to be witness to it all as #40's headlight finally popped into view.

Controlled from the watchman's tower at King Street, the wigwags at King, Waterloo, Duke, Weber and Ahrens streets activated in succession as #40 marched into town with all the pomp of an extra-fare limited. Spellbound, we watched as an olive-green Hudson strutted past in a display of white exhaust, swirling steam and clanking rods. Brake shoes grabbed hold and baggage and R.P.O. cars slid past, with baggagemen and mail clerks standing in the open doorways. Dutch doors and traps banged open as the long green coaches rolled by, and kerosene lanterns dotted the length of the train as #40 screeched to a halt at Kitchener. Dodging detraining passengers, baggage carts and mail bags, we hurried to the head end, where the massive 5700 simmered patiently, hissing steam and smelling of coal smoke and hot grease. With a friendly wave, the engineer gestured us to the cab.

Noisy, hot and dimly lit, the cab of that 5700 was at once thrilling, fascinating, overwhelming and more than a little unnerving – especially when the fireman cracked open the firebox doors to reveal a raging inferno. Like the storybook heroes, the engineer was a kindly man and he quickly put us at ease. During the short stop, he explained how to work the throttle and reverse, how the coal got to the firebox and what the gauges meant; he showed us how to blow the whistle and ring the bell; and as the highball came from the tail end, he handed us a big wad of cotton waste and a bright red bandanna.

Two short blasts on the whistle and we were back on the platform, staring up in admiration and watching intently as the engineer, silhouetted in the cab window, set the high-stepping Hudson into motion. Effortlessly, the olive-green 4-6-4 walked #40 out of town and we stood in awe until the red markers disappeared into the night. The events of that memorable evening added a new dimension to an already intense love of railroading and the inspiration for this book is rooted in the emotions stirred by that experience.

Through the formative and most impressionable years of childhood, I watched intently as steam succumbed to diesel and railroading entered an era of change that only began with the motive power revolution. Soon, passenger trains and branchlines were affected, stations, shops and roundhouses closed, and the railroad world – as I knew it – seemed to be coming apart.

The most poignant moments were the frequent visits to the scraplines at Stratford, Bright, Drumbo and London, where the steam engines that I loved languished quietly, rusting as they awaited the torch. I hoped against hope that they would somehow be reprieved, fired up and put back on the road. The reprieve never came; the lines only grew shorter as engines were dragged away to be cut up at the reclamation yard in London. At a very early age, I learned to deal with mortality.

Nevertheless, I was crestfallen the day we pulled into Stratford and discovered that the last engines were gone. Clutching my mother's Kodak Brownie, and intent upon taking my first railroad pictures, I had come to photograph the condemned giants before it was too late. The experience has left me forever sensitive to change, a champion of "survivors," and unable to merely pass by a scrapline. Instead, I find it necessary to linger in the company of the departed, take notes and photographs, reflect upon past glories and pay final respects.

Although the emotional impact of those last days of steam has never faded, my love of railroading did not die on the scraplines at Stratford and London. Indeed, the siren call of railroading grew stronger. In their own way, the green-and-gold cab units that succeeded the Hudsons and Northerns on the passenger trains through town were just as appealing, and watching CPR freights tackle Orr's Lake hill with multiples of tuscan-and-grey diesels instead of doubleheaded Pacifics and Mikes was still exciting. Diesels, I found, possessed their own personality – just as steam engines had – and I came to know the local engines by their numbers, sound and individual idiosyncrasies.

Beyond the motive power, the magic of railroading remained unchanged. Stations were still exciting places to be, coaches had the same distinctive smell, and riding the train was just as thrilling. The wonder of exploring a roundhouse or shop hadn't diminished, and station platforms, interlocking towers and yards were still good places to while away the hours. Pulses still quickened when a headlight broke over the horizon and there was still no thrill greater than when an engineer leaned down from the cab window and asked, "D'you want to come up for a ride?"

Although railroading continued to be captivating, I gave no serious thought to photography for some time after the disappointment at Stratford . . . not until a notable day in the spring of 1964. The exact date has been lost with the passage of time, but the incident remains as clear as yesterday. Interestingly, it was not sparked by any of the truly exciting moments of that year. It was not watching lightning-striped New York Central Geeps working a symbol freight through Niagara Falls, or witnessing a running meet between CNR freights on the double track at Princeton. It was not riding the CPR passenger train from Galt to Toronto, or even chasing A-B-B-A's and A-B-B's of leased Union Pacific FA's up Orr's Lake hill. Rather, the urge to pick up a camera and put railroading to film was rekindled on a quiet morning on the edge of CNR's East Yard in Kitchener.

As usual, there were very few trains around, but I became engrossed with the activities of the local yard crew as they flat-switched and kicked cars using one of the regularly assigned MLW S3's. The air was filled with the nervous-sounding, rapid exhaust of the S3's Alco 539 engine, the protesting squeal of steel wheels being forced through tight turnouts, and the shuddering grind of brakeshoes grabbing hold. Brakemen darted about, waving hand signals, pulling pins, turning switches and riding cars kicked down the ladder tracks, cranking on hand brakes, or dropping off just before the free-rolling cars crashed into standing cuts.

The high point of the exercise came as the S3 – it was 8498, I think – struggled to place a loaded Reading coal hopper on Hogg's tipple track. The slight incline was more than the 660-horsepower switcher could handle with several cars in hand and the unit stalled. I could sense the frustration and see the determination in the engineer's eyes as he took the slack and prepared to try again. There would be no humiliating double. The big man kicked off the brake and widened out on the shiny brass throttle. Sending a tall column of exhaust skyward, hood doors rattling, traction motors smoking and wheels slowly but surely turning, the little black engine shoved mightily and eventually positioned the hopper at the top of the tipple.

This, I thought, is something that must be savoured and preserved, something to be shared. That unlikely incident galvanized the commitment to put to film and to preserve the drama, excitement and the human element of railroading. Therein lies the common bond between all of the photographers whose work appears on these pages, as well as the inspiration that drove this book from a boy's dream to a reality.

From Bob Sandusky's sensitive portrayal of M391's stop at Duntroon, to Bob Gallagher's stunning rendition of CN Extra 5526 West racing ahead of a prairie thunderstorm, the mandate of this volume mirrors the motivation of the individuals whose collective efforts made this work possible. Climbing the creaky stairs of ancient interlocking towers and peering beyond roundhouse doors, exploring remote Saskatchewan branchlines and riding the Newfoundland narrow gauge, this book seeks to capture the spirit and the essence of Canadian railroading.

As a nation, Canada owes its very existence to the railroad. Prior to the construction of the Canadian Pacific Railway, Canada, as observed by Sir John A. Macdonald, constituted "little more than a geographical expression." Completion of the railway from ocean to ocean consummated the promises of Confederation and united the nation in spirit and in fact. Railway fever gripped the land and steel rails were spiked down, creating a vast network of railroads that opened new frontiers and nurtured a growing Dominion. As a nation, railroading is in our blood and the railroad has etched its signature upon the land.

The images offered herein pay homage to that legacy and present Canadian railroading in a season of transition that has been both painful and productive. During the four decades covered by this volume, Canada's railways have undergone a metamorphosis that has seen institutions topple and labour forces slashed; the sprawling network of steel that opened the land has contracted and there have been revolutions in motive power and technology. The great railway monopolies have been broken and the glory days are gone, but the magic and vitality of Canadian railroading endures.

With sensitivity and awareness, the photographers assembled here have documented the passage of the old order and they have embraced the new. They have invested much of themselves in the creation of these images, and have captured the drama and intensity of transitional railroading, as well as the poignancy of the casualties of progress. While railroading is very much dominated by machines, they have not overlooked the human element. Study carefully the people that populate these images, from the train crews and operators, levermen, labourers and shopmen going about their business, to the spectators and passengers. Give a moment to the men at Duntroon and the girls on the beach at Kelligrews; the old man and little boy strolling the platform at Barry's Bay, and the young boys waving to the circus train at Waterford. As much as anything, they are what it's all about.

The photographs contained herein are far more than celluloid images of so much smoke, steel, wood and glass; they are moments carefully and deliberately frozen in time. Come then, and experience the excitement of hooping *The Mountaineer* at Morley and the solitude of the prairie at sunset east of Arden. Stand at mileage 59 on the Galt Subdivision and thrill to the earth-shaking, ear-splitting splendour as 2220 and 5187 storm Orr's Lake hill with westbound tonnage; join the workers on the shop floor at Stratford as the 200-ton Morgan crane triumphantly cradles a freshly-shopped steam locomotive for the last time ever; venture into Coryell and revel in all the opposed-piston glory of Extra 4065 East on its knees on Farron Hill. Experience the enduring drama, the magic and the glory that was – and is – Canadian railroading.

Greg McDonnell

Orr's Lake, Ontario
March 1991

Utilizing the former Irondale, Bancroft &
Ottawa/Central Ontario Railway junction at
York River, Ontario, to turn its train, CNR
RS3 3018 shoves a combine and nine cars
around the wye before heading south with
Bancroft-Belleville mixed train M314 on
March 28, 1959.
Robert J. Sandusky

Off the Beaten Track

There was a time, not that long ago, when eastern Canada was blanketed with a far-flung network of branchlines, shortlines and interurban railroads that tapped the coal fields of Cape Breton and probed the backwoods of New Brunswick, meandered through the Quebec countryside and touched the shores of the Great Lakes. Away from the bustle of the mainline, branchline railroading was conducted with the easy pace of the local freight and the informality of the mixed train.

With frequency varying from daily to weekly, trains patrolled the thousands of miles of branchlines listed in the Canadian Official Railway Guide, pausing at team tracks and stock pens, lumber mills and Co-op elevators. Traffic was handled by the carload, or less-than-carload lot, and rural agencies were busy, if less than prosperous. It was railroading with a personal touch, from the friendly and familiar faces of the regular train crew, to the local station agent, who knew just about everyone in town. Service was personal and rarely hurried. Trains carried everything from feed grain and coal for local suppliers, to milk, mail and parcels ordered through the Eaton's catalogue. Train time was an event of local importance and the railroad was an essential part of the community.

Public timetables were thick with scheduled service to just about any point reached by rail, and train times were chalked on arrival and departure blackboards affixed to the outside walls of a thousand depots. Accommodation ranged from a dirty wooden coach tied to the tank of a Cumberland Railway & Coal 2-8-0 working #2 out of Springhill, to freshly-painted steel equipment on a Quebec Central local, or a pristine tuscan combine on the CPR Elora mixed.

Under catenary strung through the townships and rural counties of Quebec and Ontario, interurban cars provided "safe, clean, fast and efficient" passenger service to the churches and shrines at L'Ange Gardien and Ste. Anne de Beaupre; to Lake Erie beaches at Port Dover and Port Stanley; from quiet country depots at Granby, Marieville, and St. Cesaire, to the bustling streets of Montreal; and hourly cars departed Thorold for Dainsville, Humberstone and Port Colborne. Freight rolled behind zebra-striped steeplecabs or boxy freight motors, and the electric lines conducted interchange and across-the-platform connections with the "steam roads" at places like Quebec City and St. Lambert, Oshawa and Merritton, Galt, St.Thomas and London.

For better or for worse, progress has exacted a heavy toll upon the once extensive network of branchlines serving eastern Canada. Abandonments have claimed hundreds of miles of railroad. Steel gangs, rail trains and private contractors have ripped out subdivision after subdivision, and with them, the hopes and dreams of scores of on-line communities. On the surviving branches, rebuilt Geeps and RS18's perform duties once assigned to CPR D10's and ex-Grand Trunk and Canadian Northern Ten-wheelers, but LCL, stock pens, and rural agencies are extinct. Passenger service is but a memory, and even cabooses are fast becoming history. Virtually every mile of catenary has been pulled down on what remains of the interurban lines, and their corporate identities have been erased. The writing is on the wall for many more miles of branchline as the railroads continue to streamline their physical plant.

In the small towns and rural communities along these endangered lines, memories of better days linger, and there is new hope that deregulation and "spinoff" branchline sales will breathe life into their railroads. Indeed, a number of potentially viable lines have already been acquired by private interests and revived and rebuilt with the same independent spirit that inspired their construction a century or more ago.

They're still loading mail in the Model A as the 1364 gropes through the weeds at Duntroon to lift a wooden boxcar before resuming its journey to Collingwood.
Robert J. Sandusky

Hauling into Collingwood with M391, CNR 4-6-0 1364 meets sister 1355 stopped at the water plug on September 4, 1954. Still wearing their original Canadian Northern numbers, the two engines were built by MLW in 1912, and both were retired in November 1957.
Robert J. Sandusky

The fireman aboard CNR H-6-g Ten-wheeler 1364 takes a breather while the tail-end crew of mixed train M.391 handles mail during a stop at Duntroon, Ontario. The elements are timeless: the postmaster's Model-A Ford is in showroom condition and the vintage wooden boxcar in mail service is freshly shopped. However, the station has not seen a painter's brush since the Grand Trunk applied the barely visible coat of grey and green, and the weeds are journal-box high. The date is September 4, 1954, and the line through Duntroon is on the verge of abandonment.
Robert J. Sandusky

Frost is still heavy and the shadows are long as CN "RSC14's" 1777 and 1781 lead the Bridgewater-bound "Southwestern freight" past the old station at French Village, Nova Scotia, not long after sunrise on September 29, 1980. The "Southwestern" traces its name to the Halifax-Yarmouth line's origin as the Halifax & Southwestern Railway.
Greg McDonnell

Treading lightly over the spindly rails of the Middleton Subdivision, CN Extra 1801 North, the Bridgewater-Middleton way-freight, casts a perfect reflection at Union Square, Nova Scotia on May 2, 1975. One of just four RSC24's built, pug-nosed 1801 and her two surviving sisters (CN 1800 and 1803) spent most of their final years working the light-rail branches out of Bridgewater, Nova Scotia. The distinctive units were retired in 1976 and surrendered their A1A trucks to RS18's modified for lightweight branchline service.
Greg McDonnell

With heavyweight coach 1303 bringing up the markers, DAR Truro-Windsor mixed M22 ambles west of Kennetcook, Nova Scotia, on April 24, 1975. The coach will be cut off at Windsor and return to Truro on M21 later the same day, while SW1200RS 8138 and the short train will continue to Kentville.
Greg McDonnell

Homeward bound, just three cars of salt separate
RS10 8599 and steel van 434014 as engineer Don
Broadbear hurries the southbound CP Goderich
wayfreight toward West Monkton, Ontario, at
14:40 on April 17, 1978. The last CP train departed
Goderich on December 16, 1988, and the line was
officially abandoned north of Guelph on December 31
of the same year.
Greg McDonnell

A bitterly cold wind howls in off Lake Huron as CP RS10's 8464 and 8595 make up the wayfreight at Goderich, Ontario on St. Patrick's Day, 1975. There is plenty of tonnage to take south this day, and while the crew prepares to grab RS10 8586 from in front of the enginehouse and double the train together, the wind and lake-effect squalls are piling up massive drifts along the line. Towing A-rating and breaking through handrail-high drifts, the trio of RS10's will put on a memorable display working southward.
Greg McDonnell

Wet snow and freezing temperatures add to the misery as crews work to right CN GP9 4513 in the aftermath of a fatal collision between a livestock special and a gravel truck at Alma, Ontario, on November 13, 1975.
Greg McDonnell

Passing Martyr's Shrine on the entrance to
Midland, Ontario, CPR N2a 2-8-0 3632 creaks
across the Wye River trestle on March 12, 1960.
Robert J. Sandusky

Ready to bed down in the
roundhouse after a day's work,
3632 drops her fire on the
ashpit at Port McNicoll, Ontario,
on March 12, 1960.
Robert J. Sandusky

Returning from Simcoe, the CP Electric Lines "South Job" out of Preston pauses on the west leg of the TH&B wye at Waterford, Ontario, on November 4, 1981. While SW1200RS 8160 idles patiently, the LE&N crew has gathered for lunch in the 438502, conductor Bill Clack's regular van.
Greg McDonnell

Basking in the setting sun, Montreal & Southern Counties 607 and 611 rest between runs at Marieville, Quebec, on November 27, 1955. While there is no snow on the ground, the M&SC is ready for winter and the heavy snow that inevitably drifts across the Eastern Townships, as evidenced by the huge plow firmly attached to the front of 611.
Robert J. Sandusky

On a Wednesday afternoon in late March 1955, the threatened discontinuance of all CP Electric Lines passenger service is the talk of the day as GRR 626 and LE&N 937 prepare to leave Preston, Ontario, as train 214 to Hespeler. In a few short weeks, rumour will become reality, as on April 23, 1955, all passenger service on the Grand River Railway and the Lake Erie & Northern will be discontinued.
George Schaller

Working through a winter wonderland on
March 8, 1975, snow-packed CP RS10 8582
nears Keswick, New Brunswick, with
Chipman-Woodstock wayfreight #79.
Greg McDonnell

With just a single wood-chip box and van in
tow, CP RS10 8582 waits patiently with #79
as CN RS18 3890 clumps across the
diamond at South Devon, New Brunswick
on a snowy morning in March 1975.
Greg McDonnell

Train time was still a significant event in the small towns of Ontario when CPR Sharbot Lake-Renfrew mixed M783 called at Lavant on April 4, 1958. Simmering on the head end, D10g 870 waits patiently while LCL is loaded aboard the ancient wooden Baggage-RPO and locals ponder the photographer's interest in "their" train.
Robert J. Sandusky

In the final year of D4 operation out of Havelock, Angus-built D4g 484 pauses with wooden van 436822 at the CPR station in Lindsay, Ontario, on January 4, 1958. The Bobcaygeon mixed, one of the last D4 preserves, has been discontinued and the 484 is assigned to the Havelock-Lindsay wayfreight.
Robert J. Sandusky

Early on the morning of April 5, 1958, CPR D10 870 emerges from the two-stall enginehouse at Renfrew, Ontario, preparing to take M782 back to Sharbot Lake. Working the former Kingston & Pembroke, the Angus-built 4-6-0 was once assigned to, and lettered for, CP-controlled Montreal & Atlantic.
Robert J. Sandusky

Labouring upgrade out of Port Alfred, Quebec,
Roberval & Saguenay RS2 20, M420TR 26 and
RS18 25 work a train of empty hoppers back to
Arvida in April 1973. Built by MLW in December
1949, R&S 20 has been preserved and continues to
operate at the Canadian Railway museum in
Delson, Quebec.
Kenneth R. Goslett

Coal smoke drifts from the stack of the ancient wooden caboose assigned to the "Lingan Shunter" as Devco RS1 number 300 makes up its train at Glace Bay, Nova Scotia on a May 1975 day. Built in 1951 as Wisconsin Central 2366, the RS1 immigrated to Canada as Old Sydney Collieries 300 in 1961.
Stan J. Smaill

Still active at age 62, Old Sydney Collieries 2-4-0 number 25 switches hoppers for the Acadia Coal Company at Stellarton, Nova Scotia, in the bitter cold on January 24, 1962. Retired later in 1962, the venerable Baldwin has been preserved at the Canadian Railway Museum in Delson, Quebec.
Robert J. Sandusky

Ducking beneath the steelwork of the pedestrian
overpass spanning the CNR yard at Palmerston, Ontario, K-3-a
Pacific 5563 exits town with Stratford-bound train #171. On
March 30, 1957, Palmerston – the division point and hub of
CN's Bruce Peninsula branchline network – still
dispatched dozens of passenger, freight and mixed trains to
Stratford, Kincardine, Southampton, Owen Sound,
Durham and Guelph. Palmerston has lost its status as
a division point, most of the Bruce branches have been
abandoned and, in 1991 the onetime railroad town rarely even
sees a train.
Robert J. Sandusky

It's not quite business as usual as CPR D6b 526 drifts through Belwood, Ontario, on June 28, 1955. Business car 35 is tied to the tail end of the Elora mixed and today, the combine will not be coasted into the station at Elora.
Robert J. Sandusky

Light snow is falling as CP RS18's 8737 and 8765 idle away the night at Guelph, Ontario, on December 4, 1986. A fire-damaged bridge just south of town forced the Goderich wayfreight to detour over the CN and operate the temporarily isolated branch out of Guelph.
Greg McDonnell

Stopped at the home signal at Hagersville, Ontario, CN GP9's 4403, 4572 and 4518 wait with Brantford-Nanticoke #559 as Conrail train WQST-02 shoves across the diamond with a set-off on September 10, 1982. Better-known as the "Montrose Turn," the St. Thomas-Niagara Falls (Ont.) local is powered by Canadian-built Geeps 7437 and 5827, part of the fleet of GMD-built GP7 and GP9's assigned to Conrail's former New York Central/Canada Southern operations in southern Ontario. On May 1, 1985, the former Canada Southern was divided up and sold to CN and CP; the mainline has been single-tracked and the blue Geeps no longer hammer the diamond at Hagersville.
Greg McDonnell

After stopping at the Grand River Railway diamond,
CN RS18 3704 accelerates Guelph-Galt wayfreight
#580 past the nearly-abandoned Silknit textile
mill at Hespeler, Ontario, on March 17, 1987.
Greg McDonnell

Eastern Mains

While branchlines capture the sylvan charm and easy going pace of rural life, the signature of eastern railroading is cast in the 115 and 132-pound steel of heavy-duty mainlines and traced by 65-mph container trains that roll west from Halifax; by mile-long manifests dispatched from St. John and by Chicago-bound intermodal freights that sprint across Ontario farmlands toward the Detroit River Tunnel and the U.S. border; by passenger trains that top 90 mph along the Lake Ontario shore-line and transcontinental symbol freights that roll through the rocky splendour of Superior's north shore.

Steeped in history, the ancestry of these eastern mains reads like a roll call of the railways that built the nation: Grand Trunk, Intercolonial, National Transcontinental, Great Western, Credit Valley, Ontario & Quebec, Canada Southern, Canada Air Line, Canadian Northern and Canadian Pacific.

In the glory days of the steam season, railroading on the eastern mains meant CNR Northerns working heavyweight Limiteds through the Nova Scotia highlands; tripleheaded CPR P2's storming over the "Short Line" with grain trains billed to the Port of St. John, and beetle-browed CNR 4-8-4's stomping out of Turcot with westbound freights. Royal Hudsons piloted the Dominion along the north shore of Lake Superior and worked Toronto-Fort William without change, while CPR Jubilees regularly broke the century mark, working Galt Subdivision passenger runs west of Zorra. Visions of CPR G3's walking Vaudreuil-bound locals out of Montreal West in a December blizzard and "First 6" slamming through Scarborough Junction, with a CN 6200 flying green and trailing a dozen cars are indelibly etched in the memories of those fortunate enough to witness the apex of steam railroading in the East.

The heritage of the eastern mains was evident in everything from station clocks marked G.T.R., to the ex-CGR Mikes that doubleheaded tonnage over the old National Transcontinental mainline east of Monk and former-Grand Trunk USRA-design 2-8-2's that highballed time freights on the Montreal-Toronto "Double Track Route." History was kept alive in the language of train crews, dispatchers, operators and towermen. Crews at St. Thomas referred to the "Credit Main" and levermen in the interlocking tower at West Toronto lined 0-6-0 switchers onto "the old Bruce."

Two generations further removed, evidence of the old order has diminished, but the dreams that fostered the original construction of the eastern mains are alive and well and fulfilled daily. In the Maritimes, portions of the Intercolonial Railway and National Transcontinental mainlines are spiked with 132-pound welded rail and play host to hotshot container trains and manifest freights as part of CN's Halifax-Montreal artery. On the "Double Track Route" between Montreal and Toronto, stone stations of obvious Grand Trunk heritage stand in contrast to passenger trains that hurtle past at nearly 100 miles per hour and quake in the thunder of passing CN "Lasers" and doublestack container trains. On the former Great Western and Credit Valley mainlines across southern Ontario, CN/GTW and CP/SOO Montreal-Chicago runthroughs are the manifestation of century-old dreams of lucrative "overhead traffic," while on the one-time Canada Air Line, Norfolk Southern C39-8's hustle piggybacks, containers, hi-cubes and auto racks west of Fort Erie, exercising trackage rights granted by the Grand Trunk in 1898. And, in dramatic and continuing fulfilment of the greatest dream, transcontinental intermodal trains and symbol freights are dispatched daily from Montreal and Toronto, destined for the west coast on an iron road that does indeed span the nation from Atlantic tidewater to the Pacific coast.

The manifestation of Credit Valley Railway
dreams of lucrative overhead traffic, SOO/CP
Chicago-Montreal runthrough #502, lead by SOO
SD40-2 6618 and CP SD40 5401, emerges from the
"Hornby Dip", east of Milton, Ontario, at 16:50 on
October 9, 1986.
Greg McDonnell

Working eastbound tonnage over the Monk
Subdivision – one of the most spectacular pieces
of mainline railroad in eastern Canada – CN C424
3208, RS18 3831 and C630M 2030 approach Notre
Dame du Rosaire, Quebec, in June 1974.
Completion of the "Pelletier Cutoff" in 1977
diverted through traffic from this line, built by the
National Transcontinental Railway, to the former
Intercolonial mainline along the St. Lawrence River.
Kenneth R. Goslett

Early morning mist hangs heavy over the southern shore of Lac Pohenagamooke as CN Express Extra 2325 East swings through Estcourt, Quebec, in late August 1986. One of the few portions of the National Transcontinental Railway mainline still in use, the Pelletier Subdivision cuts a giant S-curve through Estcourt, barely squeezing between Lac Pohenagamooke and the Maine border.
Kenneth R. Goslett

Silhouetted in the setting sun, Ontario Northland
GP9 1602 awaits the call to duty at Englehart,
Ontario, in August 1988.
Robert E. Lambrecht

Working southbound through Cobalt, Ontario in August 1988, ONR SD40-2's 1731 and 1730 – spliced by GP38-2 1803 – pass a preserved nickel mine with mixed freight bound for North Bay.
Robert E. Lambrecht

On the last lap into the Soo, Algoma Central GP7's 170 and 100 lead Hearst-Sault Ste. Marie passenger train #2 through a blaze of fall colour on September 24, 1987.
Greg McDonnell

Soaring high above Lake Ontario, TH&B GP7's 76 and 77 pass wintering grapevines as they crest the Niagara Escarpment at Vinemount, Ontario, with the Port Maitland local on March 29, 1982.
Greg McDonnell

Clawing their way up the Niagara Escarpment, CP FP7 4036, F7B 4435, F9B 4473 and F7B 4432 grind through the S-curve at mileage 35 on the Galt Subdivision, just west of Milton, Ontario, on October 26, 1975. The annual fall movement of livestock from prairie ranches to southern Ontario pastures is in full swing and Extra 4036 West will be peddling stock at Ayr, Arkwood and other points en route to Windsor.
Greg McDonnell

Clouds of brakeshoe smoke
envelop the Paris Turn, as
non-dynamic brake-equipped
CN SD40's 5021 and 5047
descend the Niagara Escarpment
through Dundas, Ontario, on
June 11, 1972.
Greg McDonnell

Steaming in the sub-zero cold, the icy black waters of the Lachine Rapids seethe beneath the CPR St. Lawrence River bridge as D&H U23B 2305 brings the Napierville Junction local from Rouses Point, New York, into Montreal at 10:00 on January 5, 1976.
Greg McDonnell

"You are on Indian Land," proclaims the message painted on a rock nearby, and indeed, CP Extra 4501 East is in Mohawk territory as it rumbles off the St. Lawrence River bridge, over the canal lift span and past Seaway Tower at Kahnawake, Quebec on April 18, 1987. With seven units, C630's 4501 and 4502, M630's 4556 and 4509, M636 4717, and RS18's 1806 and 8752 #908's power will barely fit onto the dispatcher's train sheet, but there will be few complaints when it comes to getting the St. John-bound tonnage over the road.
Greg McDonnell

STOP, LOOK, LISTEN, warns the wigwag crossing
signal at Simcoe Street, as Conrail GP40 3006
leads Buffalo-Toronto runthrough BUCP-9 into
Niagara Falls, Ontario, at 22:05 on May 29, 1982.
Greg McDonnell

Lake Superior looms in the distance as CP C424 4248, an RS18, and an RS10 cross the Little Pic River bridge with a 48-car eastbound drag on September 26, 1970. Van Horne referred to this stretch of the CPR mainline along Superior's north shore as "two hundred miles of engineering impossibilities." Emphasizing the point, the short train simultaneously occupies a massive fill, an 819-foot trestle and a narrow ledge blasted into the rock and guarded by slide fences.
James A. Brown

Heading into the horseshoe curve skirting Jack Fish Bay on the north shore of Lake Superior, freshly painted CP Rail M636 4725 exits Jack Fish Tunnel leading SD40 5535, sister M636 4723 and another SD40 on an eastbound symbol freight. In May 1885, the last spike on the Montreal-Winnipeg mainline was driven just west of here, at what is now mileage 102.7 on the Heron Bay Subdivision. *Kenneth R. Goslett*

Ancient timbers groan in protest as CP RS18 8776,
FB2 4468 and RS3 8426 lead eastbound tonnage
across the wooden trestle at Magog, Quebec, on
September 17, 1973.
John Sutherland

The staccato bark of Alco 251's shatters the calm of an August 1989 evening near St. Eleuthere, Quebec, as CN M420's 3536, 3527, 3572 and HR616 2103 drag a westbound freight out of the Riviere St. Francois valley and away from Lac Pohenagamooke, still visible in the distance.
Kenneth R. Goslett

Working northbound ore empties on June 8, 1981, Quebec Cartier Mining M636 82 and C636 77 (ex-Alco demonstrator 636-3) follow the Riviere aux Rochers north of Port Cartier, Quebec. Unseen, but shoving hard on the tail end, is radio-controlled M636 helper engine 44.
Greg McDonnell

Assigned to the Lambton assist pool – an ignominious end to a distinguished career – CPR Royal Hudson 2839 leads RS18 8795 up Orr's Lake hill, west of Galt, Ontario, on May 16, 1959.
Paterson-George Collection

Slogging up Orr's Lake hill on June 8, 1957, CPR G1 Pacific 2220 and P1e Mikado 5187 hammer past the east mileboard Orr's Lake and into Barry's Cut with westbound tonnage. After assisting the 5187 for the 55-mile run from Lambton Yard in Toronto, the 2220 will cut off at Orr's Lake, three miles west of Galt, Ontario, and return light, while 5187 will continue with its train to London.
Robert J. Sandusky

Late in the evening of March 5, 1960, CPR H1c
Royal Hudson 2839 pauses at Guelph Junction,
Ontario, after assisting a single FA on a westbound
freight out of Lambton. One of the last Royal
Hudsons to operate, 2839 was saved from the torch
by the Ontario Government and sold to a group of
Americans after deteriorating in outdoor storage
for many years. A rusting, vandalized hulk, 2839
was moved to Pennsylvania in 1972 and given a
complete overhaul. In 1979, the reborn Royal
Hudson was leased to the Southern Railway for
excursion service, taking the former-CPR engine
to destinations not even dreamed of on that March
night in 1960.
Robert J. Sandusky

Accelerating out of St. Clet, Quebec in April 1971, CP Extra 4050 West, #903's freight, heads for Smiths Falls in a blaze of Alco glory after stopping to comply with the 40-mile train inspection specified by timetable "special instruction C."
Stan J. Smaill

Dwarfed by the imposing form of Mont Orford, CP C424's 4245, 4200 and 4207 – spliced by RS18's 8749 and 8783 – skirt the frozen shore of Orford Lake with St. John-bound tonnage west of Mont Orford, Quebec, in early April 1977.
Kenneth R. Goslett

With just six months left to live, CN CFA16-4 9306 slams through Lorne Park, Ontario, leading RS18's 3848 and 3672 on Extra 9306 East on May 28, 1966.
James A. Brown

Flying freshly laundered white flags, World War I-vintage CNR S-3-a Mikado 3739 storms through Dorval, Quebec, with an Extra West on March 23, 1957. Built as Grand Trunk 479, the 3739 was one of 40 USRA-design Mikes delivered to GTR by Alco in 1918. *John Welsh (Carleton Smith Collection)*

Seen from the cab of CN FP9 6536, there are "highballs all around" as the westbound *Lakeshore Express* slams past CN Extra 9434 East and overtakes a westbound drag in the hole at Prescott, Ontario, on July 7, 1962. *James A. Brown*

Working a Windsor-Buffalo hotshot, and exercising trackage rights over CN's former-Canada Air Line route, a trio of Canadian-built Wabash F7's slam past CN Extra 3875 West, "high and dry" in the siding at Canfield Junction, Ontario, on May 28, 1966.
John Freyseng

Just about to hit the diamond crossing the CPR North Toronto Subdivision, CNR U-2-g Northern 6233 passes the West Toronto interlocking tower with Sarnia-bound train #11 on August 8, 1958.
Robert J. Sandusky

Inside West Toronto Tower, the plant is lit up like a Christmas tree as leverman James A. Brown takes care of business on an August 1960 night.
James A. Brown

Lightning-striped New York Central PA1 4208 and black E8A 4052 bring a westbound mail train into Windsor, Ontario, in April 1962, while Canadian-built NYC GP7 5825 waits in the wings.
L. Norman Herbert

Just out of Montreal West, Quebec, D&H PA1's 18 and 17 shrug off the snow as they drop down the "South Junction Lead" with Montreal-New York train #34 The Laurentian on February 19, 1971. Exercising trackage rights over the CPR, the handsome Alco cabs will regain home rails at the Delson, Quebec junction with D&H subsidiary Napierville Junction.
Greg McDonnell

As a vintage BA Oil Co. "Super Duty" Ford rumbles overhead, CP Extra 4002 West bursts out from under the Highway 401 overpass west of Toronto Yard in Scarborough, Ontario, in February, 1965. By fall, Alco-built FA1 4002 and MLW-built sister 4011 will be retired and traded in on C424's, while FB1 4408 will soldier on until May 1975.
John Freyseng

It's a clear board at "JC" as leased Union Pacific FA1 1627 and two UP FB1's work a westbound CPR freight through Streetsville, Ontario, on the evening of March 30, 1964. The three Alcos were among seven UP FA1's and eight UP FB1's leased to CP during the 1964 grain rush.
James A. Brown

At 18:20 on the sweltering-hot evening of August 16, 1987, two boys watch from the milldam as CP Chicago-Montreal runthrough #502 crosses the Grand River bridge at Galt, Ontario, with three deadheading cabooses cut in ahead of the "working van."
Greg McDonnell

In the last light of an August 1975 evening, CN FPA4 6761, F9B 6616 and RS18's 3120 and 3123 rumble over Riviere Becancour at Daveluyville, Quebec, with the eastbound Montreal-Halifax *Ocean*. *Kenneth R. Goslett*

With a friendly wave, Glen and Ted McDonnell greet the "Greatest Show on Earth," as B&O GP40 4058 and C&O GP7 5736 wheel the Ringling Bros. Barnum & Bailey Circus train eastbound through Waterford, Ontario, at 12:15 on October 12, 1981. Ducking beneath the long-out-of-service Lake Erie & Northern overpass, the C&O train is exercising trackage rights over Conrail's ex-Canada Southern mainline between St. Thomas and Niagara Falls. *Greg McDonnell*

CP Rail Extra 4742 East waits patiently in the siding at Blandford, Ontario, as the head-end brakeman cleans out the switch to allow #70's freight out onto the mainline after a meet with "Make-up 903" on January 10, 1987. *Greg McDonnell*

With Alco 251's shouting in competition with the crashing thunder of a violent thunderstorm, CP "Big Alcos" 4722, 4562 and 4734, along with C424 4214, slog through Drumbo, Ontario with the "Putnam Potash" on June 22, 1987. After being stopped at Wolverton due to water over the rails, engineer Jerry Chewhowie has his work cut out for him, getting more than 10,000 tons rolling on wet rail and a slight upgrade. *Greg McDonnell*

For a brief moment at sunset, the sun peers from behind heavy black clouds to illuminate CP RS18 8790 and two M636's working #904 across the Grand River bridge at Galt, Ontario, on October 16, 1973. *Greg McDonnell*

On a pleasant day in September 1988, engineer Joe Hickey eases Terra Transport NF210's 933 and 915 past the scale house at Corner Brook with #203's freight. Part of the true intermodal service provided by Terra Transport, the green containers on the head-end represent the future of Newfoundland railroading. Appearances are deceiving though, for Hickey, the 933, 915, train 203 and indeed the entire railroad are operating on borrowed time. Within a month, Terra Transport will cease all regular rail operations, the containers will cross the island by truck, and Canada's tenth province will be without a railroad. *Stan J. Smaill*

Island Railroading

For most of their existence, the railroads of Canada's two island provinces had precious little in common. Although both systems were constructed as 42-inch narrow-gauge railroads that became wards of the federal government with Confederation, the history, character and operations of the Prince Edward Island Railway and the Newfoundland Railway were as different as the islands themselves.

Once labelled "the worst-built railway ever to be seen in North America," the PEIR meandered easily through the lush, rolling terrain of the picturesque island. Despite takeover by the CNR, conversion to standard gauge and connection to the mainland (via car ferries plying Northumberland Strait between Borden and Cape Tormentine, New Brunswick), PEI railroading retained the pastoral charm of its environment.

In the diesel era, GE 70-tonners, CLC H12-44's, A1A-trucked MLW roadswitchers and reefer-loads of potatoes were staples of the 274.7-mile island railroad, but the atmosphere was little changed from the post-World War I days of Ten-wheelers and dual-gauge track. The urgency, and solvency, of mainline railroading remained forever foreign to PEI soil.

Statistically at least, Newfoundland possessed all the trappings of mainline railroading: a 547.8-mile mainline, division points and junctions with connecting branch-lines, trans-insular passenger service, mixed trains and timetabled "fast freights." In the flesh, however, Newfoundland railroading was an experience unique unto itself.

By necessity, the railroad retained its original 42-inch gauge and with Newfoundland's entry into Confederation on April 1, 1949, CNR inherited a 704.3-mile narrow-gauge system stretching from Port aux Basques to St. John's, complete with branchlines to Carbonear, Placentia and Argentia, Bonavista and Lewisporte. In the steam season, railroading in Canada's tenth province was synonymous with doubleheaded narrow-gauge Mikados hammering across the desolate Northern Barrens with the Caribou and dapper Pacifics, outside-braced boxcars and open-end wooden coaches tracing the rocky shoreline of Trinity Bay.

The custom-built 900 series GMD NF110's, NF210's and 800 series export-model G8's that bumped steam in the mid-1950s, became fixtures of Newfoundland railroading for more than 30 years. Calling on quiet fishing villages, ports and lumber camps, trios of the boxy 800's worked branchline mixed trains, while the 900's ruled the mainline. Out on the Barrens and deep in the boreal forest, along the rocky coastline and in villages, division points, junctions and towns from Port aux Basques to St. John's, the chant of GM 567's, the call of three-chime air horns and the sight of multiples of 900's working 80-car freights were the essence of mainline narrow-gauge railroading.

In an economic climate as harsh as the Newfoundland winter, the narrow-gauge fought a never-ending battle for survival, establishing "interchange" with the mainland railroads (by re-trucking standard gauge cars at Port aux Basques), sacrificing the cherished "Newfie Bullet" and introducing true intermodal container service. The Newfoundland narrow-gauge fought the good fight in its effort to survive, but in 1988 the railroad was sold out in return for $800 million in highway funds.

On September 30, 1988, regular rail service on Newfoundland came to an end as the mixed train from Bishop's Falls pulled into Corner Brook. Fifteen months later, CN quit Prince Edward Island and the railroads of Canada's two island provinces shared a common, and unfortunate, fate. The last trains have been OS'ed out of Bishop's Falls and Corner Brook, Borden, Tignish and Royalty Junction; the chant of labouring 567's no longer drifts across the Barrens and the rails are ripped up through St. Andrew's and St. Fintans, Harry's Brook and Corner Brook, Millertown Junction and Arnold's Cove. On Prince Edward Island and Newfoundland, the trains are gone for good.

Onlookers take in the action on the platform at St. John's, Newfoundland, as doubleheaded CNR narrow-gauge Mikados 323 and 314 are readied to start train #1 on its trans-island trek to Port aux Basques on June 22, 1956. Less than ten years old, the MLW-built, ex-Newfoundland Railway 2-8-2's are making their last stand. Nine GMD NF110's are already on the island, and even as the young Mikes prepare to exit St. John's with #1, workers at the General Motors Diesel plant in London, Ontario, are assembling six-motor NF210's that will forever banish steam from the former Newfoundland Railway.
Robert J. Sandusky

Still dressed in the green and gold of yesteryear, CNR G8 803 rests in front of the Newfoundland Railway shops in Saint John's with sister 802 on a Sunday morning in May 1975. Come Monday, the two G8's will be bound for Argentia and the Bonavista branch. *Stan J. Smaill*

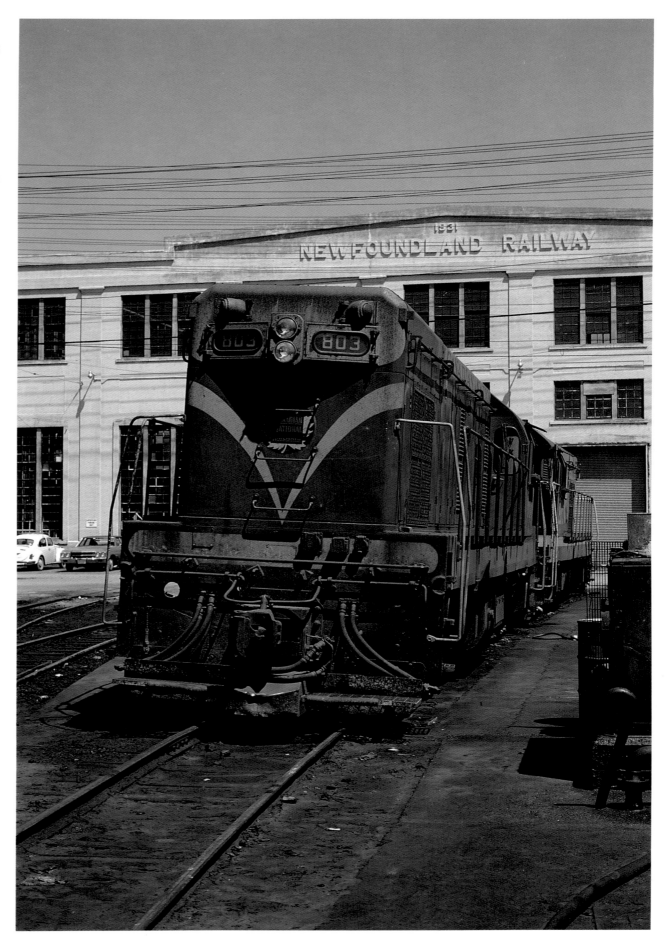

Tracing the rocky shoreline of Conception Bay, a trio of CNR G8's lead the Carbonear mixed through Kelligrews, Newfoundland on June 22, 1967.
James A. Brown

Westbound #401 is waiting in the siding, and the conductor and tail end brakeman are out on the rear platform of CNR "comboose" 6009, a converted Newfoundland Railway second-class coach, as #208, the Carbonear mixed rolls through Kelligrews on June 22, 1967.
James A. Brown

East of Port aux Basques, CN NF110 913 and sister
NF210's 946 and 933 hustle trans-island freight
204 along the rugged Newfoundland coastline in
May 1975.
Stan J. Smaill

Two days before its last run, train #102 the
Caribou snakes through the rough terrain near
Wreck House, Newfoundland, on June 29, 1967.
The fireman's mission on the gangway of NF110
907 is unknown, but up in the cab, engineer David
Dicks is getting plenty of help from his two sons.
The Dicks family consisted of three generations of
railroaders, serving the narrow-gauge system to the
end, even assisting with dismantling operations.
Discontinuance of the legendary "Newfie Bullet"
on July 2, 1967, marked the beginning of the end
for the Newfoundland narrow-gauge.
Robert J. Sandusky

Lead by a trio of 800-class G8's, the Tuesday, Thursday and Saturday-only "Carbonear mixed," train 211, drops into Carbonear, Newfoundland, on June 22, 1967. Passenger accommodation is provided in the forward section of ramshackle "comboose" 6009 tied to the tail-end of the train. Stowed in the baggage section of the car, the train crew's fishing gear affords a subtle indicator of the tiny train's adherence to schedule.
James A. Brown

Working amid lobster pots and fishing boats, three of CN's last four GE 70-tonners switch the pier at Souris, Prince Edward Island, on April 28, 1975. All four surviving CN 70-tonners were assigned to Charlottetown, and the light-rail branches on the east end of the island remained their domain until retirement. Number 30 has been preserved and continues to operate at the Canadian Railway Museum in Delson, Quebec.
Greg McDonnell

Not long after dawn, CN RSC13's 1732 and
1702 work train 550 through the fog at
Alberton, Prince Edward Island, on
April 29, 1975. Between Tignish and Summerside,
the train will stop at nearly every town and lift
potatoes loaded in aging MDT and NRC reefers.
Greg McDonnell

Returning from Souris, CN Extra 30 West
prepares to enter the Borden Subdivision at
Royalty Junction on April 29, 1975. Having
set off four loads of potatoes for the boat
train, the local is returning to Charlottetown
with just two cars and a van.
Greg J. McDonnell

With MacIntosh & Seymour 539's wound full out, CN RSC13's 1727, 1730, 1734, 1732 and 1702 put on an unforgettable display as Extra 1727 East storms out of Kensington, Prince Edward Island, Borden-bound with full tonnage on April 29, 1975.
Greg McDonnell

Bound for Borden and connection with the M.V. *Abegweit*, CN H12-44 1639 hurries Charlottetown-Moncton #39 through Kinkora, Prince Edward Island, on January 26, 1962. At Borden, #39's consist will be loaded aboard the *Abegweit* for the journey across Northumberland Strait to Cape Tormentine, while the 1639 will remain on the island.
Robert J. Sandusky

The two-storey station at Mt. Stewart Junction, Prince Edward Island, hosts CNR "winter-only" mixed trains M233 and M249 on January 25, 1962. With a single CPR boxcar, a wood-sheathed baggage car and coach in tow, GE 70-tonner 42 will work M233 to Souris, while sister 29 and steel combine 7175 make up M249 bound for Georgetown.
Robert J. Sandusky

Yard Limits

If the spirit of railroading is best manifested on the high iron, where the long freights and limiteds roll, then surely its heart beats within the yard limits. Between the yellow, diamond-shaped Yard Limit signs, the drama of the mainline melds with a more intimate style of railroading, where the human element is highly visible and many of the secrets and mysteries of the railroad are revealed.

Beneath the semaphore blades of train order signals, behind doors marked "Conductors Only," and in stations, yard offices and dispatching offices, generations have been made privy to the intimacies of railroading. In dispatching offices, all the drama, intensity and action taking place on hundreds of miles of railroad is simultaneously concentrated in a small, often tension-wracked room. In the very nerve centre of the railroad, dispatchers unravel the mysteries of CTC and the lost art of train-order railroading, or rattle off "OCS clearances" via radio and microwave relay to trains just outside or hundreds of miles distant.

In yard offices and crew rooms, work is lined up and switching manoeuvres are plotted. Train crews book on, and off, and around undesired turns; tales of railroading past and present are relived and the human side of the railroad is manifested in the personalities, faces and names that are the heart and soul of railroading.

While lacking some of the glamour of the mainline, railroading within the yard limits is, in its own way, just as impressive. Long freights and diminutive locals roll in off the mainline, lift and set off cars, change crews and continue on or terminate. Yard engines squeal through tight-curved industrial spurs, trailing cars for local industries, or saw back and forth, flat-switching the yard and kicking cars, lining up the lift for the nightly through freight, or making up a train for the local wayfreight. Brakemen cling to the sides of cars, and drop off at speed, pull pins, make joints and throw switches. Lanterns bob in the dark, swinging hand signals and illuminating car numbers. It's a world of radio chatter and switch lists, waybills and journals; it's squealing wheels, crashing couplers, barking train masters and coffee brewing in the van. It's the gritty, bare-bones essentials of basic railroading. The railroad bares its soul within yard limits, affording an intimacy and accessability available nowhere else.

Smoke from the car cleaners' fires drifts across the CPR yard in Nelson British Columbia, as power for the Midway Turn – CPA16-4 4053, H16-44's 8709, 8724 and CPA16-4 4104 – leaves the shop at 10:00 on April 9, 1974. The Turn's head-end brakeman is walking ahead to get the switch, while three tracks over, the car-knocker inspects the train. Opposite the two wooden vans just ahead of the 4053, the Nelson yard crew discusses their switch list, while in the distance, the Nelson Auxiliary, complete with steam crane 414325, slumbers quietly near the diesel shop. Examine carefully the intimate details of John Sutherland's overhead view of the Kootenay Division terminal, for it presents a classic study in yard-limits railroading.
John Sutherland

Pigeons flap overhead as venerable CP S2 7055 rumbles into Winnipeg Yard with a cut of cars from Weston Shops on February 2, 1981. *Greg McDonnell*

In the bitter cold of a February 1981 morning, pigeons, in search of spilled grain and radiant heat, gather on top of grain hoppers in the CPR yard in Winnipeg, Manitoba. *Greg McDonnell*

The engineer of a homeward-bound CPR local is
taking maximum advantage of 6603's 60-mph
gearing as the 20-year-old S10 screams eastward
through Mimico at 20:30 on July 31, 1978. Exercising
CP trackage rights on the CN Oakville Subdivision,
the short train will regain home rails at Cabin D
in downtown Toronto and tie up for the night at
Parkdale Yard.
Greg McDonnell

Backing toward a wooden boxcar of similar vintage,
1903-built CNR 0-6-0 7222 switches the car shop
behind Spadina roundhouse in Toronto, Ontario,
in the summer of 1954. Built by company shops as
Grand Trunk 1654, the diminutive switcher was
retired in July 1955, at age 52.
James A. Brown

Dwarf signals and distant fires twinkle in the sunset as TTR sectionmen use kerosene to burn the ice from switch points at the west end of Toronto Union Station on February 6, 1982. *Greg McDonnell*

Switch tenders at Cabin D, just west of Bathurst Street in Toronto, are being kept busy on a February 1966 day, as CN FPA4 6791 heads eastward toward Union Station with a single heavyweight coach and S4 8175 switches the coach yard lead. In the distance, a lone CPR RS10 waits for the light at CP's Tecumseh Street tower, while a CN S2 trundles toward Parkdale. *John Freyseng*

The palpitating beat of a 6-cylinder MacIntosh & Seymour 539 engine fills the night air as CN S7 8232 idles in front of the station at Kitchener, Ontario, on February 5, 1975. As evidenced by the blurred light of a brakeman's lantern at the station door, the 8232 is about to resume her nightly chores. *Greg McDonnell*

One of only two diesels on the roster of the otherwise all-electric London & Port Stanley Railway, London-built G12 L5 switches the L&PS freighthouse in London, Ontario, in September 1964. The L&PS electrificiation was de-energized upon the CN takeover of the line on January 1, 1966, and L&PS G12's L4 and L5 ended their days on Vancouver Island as CN 991 and 992.
L. Norman Herbert

On a hot August 2, 1967, the CPR yard crew at Goderich, Ontario, takes a break on the platform, while number 17, their Caterpillar-powered, CLC-built switcher purrs contentedly nearby. The Goderich yard job was long ago abolished and the yard was torn out following the abandonment of the line north of Guelph in 1988. The station, however, has been preserved and CP 17, sold to Babcox-Wilcox in Galt in 1974 awaits possible preservation as well.
George W. Roth

Dwarfed by the N.M. Paterson & Sons bulk
carrier *Mantadoc*, CN SW900 7921 rumbles
along the dock after placing a pair of covered
hoppers at the Canada Malting elevator on
the Toronto waterfront on December 23, 1986.
Greg McDonnell

With a sharp eye for hand signals and a tight grip
on the brake, the engineer on CP S4 7108 gingerly
shoves a cut of Vancouver Island-bound cars
aboard the CP car ferry Princess of Vancouver at
Vancouver, British Columbia, on May 14, 1967.
John Freyseng

CN "RSC14" 1759 positions a string of brand-new
tank cars – loaded on CN flats – on Pier 30 in
Halifax, Nova Scotia, in November 1988. Built for the
Zaire Railways by National Steel Car in Hamilton,
Ontario, the cars will be loaded aboard the waiting
freighter *Silver Gulf* for shipment to Africa.
Doug Conrad

CPR G5c 1270 simmers just outside the bay window with an excursion train as the operator at Ste. Agathe, Quebec, copies train orders for the passenger extra's return trip to Montreal on October 16, 1960.
Don McQueen

Thanks to the "Hi-Speed Delivery Fork Company" of Shelbyville, Indiana, the days of wooden train order hoops are gone, and operator Norm Farmer's delivery of orders to the engineer of CP Extra 8594 West at Guelph Junction at 15:49 on September 5, 1976, is painless.
Greg McDonnell

"The order board's on at GU" and operator Murray Dronick is out with the hoop as the CP "London Pick-up" approaches the station at Guelph Junction, Ontario, with RS18's 8782, 8783 and SW1200RS 8156 at 17:00 on July 2, 1977.
Greg McDonnell

The conductor on the van of CP Extra 5005 West simultaneously tosses off the waybills for the set-off and picks up orders from operator George McDonnell as the "London Pick-up" exits Galt, Ontario in a hurry on November 3, 1979.
Greg McDonnell

Locals and Limiteds

For generations of Canadians, travel by train has been synonymous with images of the glory days of railroading. Images of quaint frame depots and great steel limiteds, mixed trains and backwoods locals, trains with grand-sounding names like *Imperial* and *Overseas Limited*, *Maritime Express*, *Dominion* and *Canadian*, and trains that rated no more than numbers. Images of plush-upholstered walk-over seats and white-haired conductors in brass-buttoned vests. The sound of hissing steam and smell of brakeshoes at station stops, the crash of step boxes being heaved aboard, the bang of traps and Dutch doors slamming shut, the conductor's call and the lurch as the train gets under way. Images of the "organized chaos" of a busy dining-car galley and the soul-satisfying serenity of drifting to sleep between crisp white sheets and warm woollen blankets, listening to the rhythm of the wheels and watching the country slide past a roomette window. Images that were almost universal, whether rocking through Gaff Topsail or Codroy Pond on the "Newfie Bullet," or twisting through the Spiral Tunnels in the stainless-steel splendour of *The Canadian*. From Argentia, Almonte and Alliston, to White Horse, Xena, Yoho and Zealandia, passenger trains have been an essential element of the Canadian experience, uniting the nation in spirit and in fact.

The automobile, the paved highway, a changing society and the evolution of railroading have inflicted hard times upon the Canadian passenger train. Throughout the sixties, seventies and eighties, train-off petitions and notices of discontinuance were plastered on waiting-room walls, at ticket wickets and on passenger car bulkheads across the land. Small-town stations were boarded up and bulldozed by the hundreds, branchlines were torn up, and the mixed and local passenger train was pushed to the brink of extinction. One by one, even the mighty fell, as celebrated name trains got the axe – the *Dominion, Overseas* and *International, Mountaineer, Panorama* and *Washingtonian*. In one fell swoop, the infamous VIA cuts implemented on January 15, 1990, slashed the national rail passenger system almost in half. Among the victims was *The Canadian*, once the flagship of "The World's Greatest Travel System" and unquestionably Canada's most treasured train.

Images of the Canadian passenger train in this final decade of the twentieth century form an enchanting mosaic, a mix of old order and new, set against the contrasts and contours of the land itself. VIA LRC's hurtle along the Quebec-Windsor corridor, where the promise of true high-speed rail threatens to become reality, while mixed trains tread carefully over the muskeg and tundra of northern Manitoba. World War I-vintage box-cab electrics grind out of Montreal towing heavyweight coaches crowded with weekday commuters, while GO Transit F59's whip bi-level trainloads of their Toronto counterparts past QEW traffic jams at 60, 70 and 80 miles per hour.

It is still possible to ride to the edge of James Bay behind Ontario Northland Geeps and through the Coast Mountains and the wilds of the British Columbia interior aboard BC Rail RDC's. Algoma Central dispatches Geep-hauled passenger trains from Sault Ste. Marie to the Agawa Canyon and through to Hearst, while Quebec North Shore & Labrador SD40's handle bi-weekly passengers from Sept Iles to Labrador City and Schefferville. It is still possible to book sleeping-car space out of Churchill and watch the northern lights from the window of a VIA bedroom; to ride the ex-Grand Trunk Pacific to Prince Rupert, the Hervey-Cochrane fragment of the old National Transcontinental mainline, and to end of steel at Gaspe.

As Joseph Howe dreamed in 1851, it is also possible "to make the journey from Halifax to the Pacific in five or six days."

Despite train-offs and abandonments, nationalization and rationalization, Canada's passenger trains remain very much a part of the Canadian experience and cling tenaciously to their role as symbols of national unity.

In a spectacular display of rich tuscan and glistening stainless steel, the transcontinental flagships of the "World's Greatest Travel System" meet at Canmore, Alberta on November 6, 1960. Lead by FP7 1404, #1 has put away 2,307 miles of its 2,881.3 mile, 71-hour Montreal-Vancouver trek. CPR #2, meanwhile, has logged only 547 miles and will meet three more editions of #1 before arriving at Windsor Station in Montreal in two days.
Robert J. Sandusky

They're sipping cognac in air-conditioned comfort back in the club car, but on the head-end, the cab windows are down and the nose and cab doors are propped open in deference to the sweltering heat as CN FPA2u 6759 and FPB4 6864 approach Kingston, Ontario, with Montreal-Toronto train 55 on June 29, 1975.
Greg McDonnell

Just in from Ottawa with Pool Train #263 on
April 5, 1958, CPR G5c 1267 waits while passengers
detrain behind the CNR station at Brockville, Ontario.
Included in #263's consist are Ottawa-Toronto
coaches and a parlour car that will be added to
Montreal-Toronto Pool Train #15.
Robert J. Sandusky

CN FP9A 6502 leads a matched A-B-B-A of GM cabs
and 19 cars along the banks of the Fraser River as
CN #2, the eastbound *Super Continental* clears
Cheamview, British Columbia, on June 19, 1970.
James A. Brown

In the time-honoured tradition, a CPR trainman welcomes a passenger aboard #11 as the Toronto section of *The Canadian* prepares to depart Toronto Union Station on October 25, 1977.
Greg McDonnell

The short consist of the eastbound *Canadian* casts a long shadow at Lake Louise, Alberta, just seconds before ducking beneath the Trans-Canada Highway on January 22, 1976. With noted CPR engineer Floyd Yeats at the throttle and FP9A 1414 and freshly painted "torpedo" GP9 8517 in charge of the seasonal minimum of just seven cars, #2 will make Calgary in good time.
Richard Yaremko

CN sleeper *Eldorado* carries kerosene marker
lamps, as well as a back-up light, as the consist of
VIA #94, the *Hudson Bay* basks in the crimson
glow of a sub-arctic sunset at Churchill, Manitoba,
at 16:20 on January 30, 1981.
Greg McDonnell

It's homemade soup and lasagna for supper
as the chef on the *Hudson Bay* works in the
galley of ex-CN dining car 1346 on
January 30, 1981.
Greg McDonnell

The northern lights are shimmering overhead as
#94 pauses at Gillam, Manitoba – the first division
point south of Churchill – at 01:30 on January 30,
1981. The consist behind CN F7Au's 9155 and
9151 includes two VIA steam-generator cars and
four steam line – equipped mechanical refrigerator
cars, as well as a mail car, baggage car, a pair of
coaches, diner 1346 and sleeping car *Eldorado*.
Greg McDonnell

Less than two years old, but already looking some-
what worse for wear, CN H12-64 7624 departs St.
Lambert, Quebec, on August 2, 1954, with train
706. The Montreal-Waterloo local will leave the
CN at M&SC Junction and continue southward
on the interurban trackage of the Montreal &
Southern Counties.
Robert J. Sandusky

From the train orders neatly tucked between the air lines, to the massive "Weston Electrical Instrument Company" ammeter and the polished brass throttle, the cab interior of 1914-vintage CNR Z-1-a box-cab electric 6713 shows little evidence of modernization at Central Station in Montreal on March 7, 1975.
Greg McDonnell

About to enter the Mount Royal Tunnel, CN Z-4a electrics 6722 and 6724 brake into Portal Heights with morning commuter train #956 on May 28, 1975. Younger sisters of the venerable Z-1's, the nine CN Z-4's were built by English Electric for the Montreal Harbour Commission Terminal Railway between 1924 and 1926, and traded to CN in 1940 in return for 0-6-0 switchers 7512-7518. Five of the English Electrics were retired in 1971, but the remainder survive in Montreal commuter service in 1991.
Greg McDonnell

Cutting a swath through the ice-packed Baie de Gaspe, a freighter slowly passes as VIA FPA4 6761 drifts through Douglastown, Quebec, with #17, the *Chaleur* on March 22, 1989. In nine days, federal regulations requiring "RSC" safety control equipment on all lead units will banish the famed Alco cabs from this and almost all other VIA trains, and by spring, all of the FPA4's will be retired.
Robert E. Lambrecht

Bucking a late-season blizzard in April 1975, CN FPA4 6772 leads #12, the *Scotian*, through Routhierville, Quebec, with a perfectly matched, all-Alco A-B-A.
Stan J. Smaill

Running out of a blinding snow squall, snow-packed VIA FPA4 6764 hurtles Sarnia-Toronto train #84 through Baden, Ontario, on February 13, 1988.
George W. Roth

Passengers detrain from BC Rail RDC1's BC-21 and BC-22 as the operator at Pemberton, British Columbia, waits to deliver train orders to the conductor of North Vancouver-Prince George train #1 on January 27, 1990. Acquired from SEPTA, the two BCR RDC's were originally Reading 9156 and 9160.
Greg McDonnell

In the Land of Evangeline, CP RDC1 9067 drifts across the Moose River bridge at Clementsport, Nova Scotia, with Dominion Atlantic #1 on April 25, 1975. Discontinuance of passenger service on the DAR on January 15, 1990, sounded the death knell for most of the railroad and the trackage west of Kentville, 140 miles to Yarmouth, was abandoned on March 27, 1990.
Greg McDonnell

Assigned to the shuttle between Guelph and Guelph Junction, CPR Electro-Motive/St. Louis Car gas-electric 9004 rests in front of the station at Guelph, Ontario on September 10, 1956. Connecting with mainline passenger trains at Guelph Junction, the zebra-striped car made the 15-mile trip up to 12 times per day.
John Welsh (Carleton Smith Collection)

Rounding the wye with "daily except Sunday" train 89 from Ottawa, CNR K-3-b Pacific 5583 clanks past the water tank at Barry's Bay, Ontario, on April 5, 1958. Through service to Depot Harbour ended in 1933 when a washout severed in the former Ottawa, Arnprior & Parry Sound line at the Cache-Two Rivers trestle in Algonquin Park. The train will be wyed and return to Ottawa as #90. The combine parked by the enginehouse provides passenger accommodation on the Tuesday and Friday mixed train between Barry's Bay and Whitney.
Robert J. Sandusky

Seventy-five miles and nearly three hours out of Barry's Bay on April 5, 1958, #90 pauses at Galetta, Ontario, while the 5583 slakes her thirst at the steel water tank.
Robert J. Sandusky

Ready to return to Ottawa, CNR #90 awaits departure time in front of the station at Barry's Bay,
Ontario on April 5, 1958. Waiting for the appointed
moment, an old man and little boy stroll the platform toward the simmering, ex-Grand Trunk Pacific.
Passenger service to Barry's Bay was discontinued
in 1963 and the line was abandoned in the late
1980s. The old OA&PS station, however, survives
as a senior citizens' centre.
Robert J. Sandusky

The order board is on and operator Vic Burton,
hoops in hand, is in position as #14, the eastbound
Mountaineer, cruises through Morley, Alberta,
behind a trio of CPR GP9's on August 15, 1960. A
CPR/SOO Line pool train, the *Mountaineer* operated
between St. Paul, Minnesota and Vancouver, British
Columbia, from late June until the end of August.
During the rest of the year, St. Paul-Vancouver
service was timetabled as the *Soo-Dominion* and
involved a connection at Moose Jaw between the
transcontinental *Dominion* and SOO/CP St. Paul-
Moose Jaw trains 13 and 14.
Robert J. Sandusky

Already relieved of their orders, the wooden train-order hoops have been tossed back onto the platform as #14 clears Morley, Alberta, on August 15, 1960. With the traditional open observation car bringing up the markers and St. Paul-bound Pullmans tucked in ahead, there is no mistaking the *Mountaineer*. While other CPR name trains through the mountains rated stainless-steel domes by this time, the *Mountaineer* upheld a long-standing summer tradition and carried an open observation car west of Calgary.
Robert J. Sandusky

Bracketing RS18 3111, green, black and gold CNR FPA4's 6782 and 6788 hurry train 211 through Saguenay Power, Quebec, on August 30, 1962. Head-end equipment, an ancient wooden baggage-express car and four newer, steel sisters outnumber passenger cars on the Chicoutimi-bound train.
Robert J. Sandusky

Wearing fading but original paint, CN FP9A 6541 and an F9B freshly painted in the new colours lead #52, the Toronto section of the *Super Continental*, through a rock cut near Black Road, Ontario, on June 1, 1962.
James A. Brown

Stepping lively after an all-night run from
Regina, freshly overhauled CNR K-3-g 4-6-2
5626 brings #7, *The Owl* into Saskatoon,
Saskatchewan, on July 7, 1959. The highest-
numbered CNR Pacific, 5626 was originally
Grand Trunk Pacific 1114, the last of 15
Pacifics built for GTP by MLW in 1911.
Paterson-George Collection

The baggagemen are waiting with carts piled high with mail bags and luggage as CP FPA2 4096 and RS10 8470 bring #15, the Toronto section of the Vancouver-bound *Expo Limited* into Sudbury, Ontario, on May 26, 1967. The Toronto and Montreal sections of the westbound *Expo Limited* will be combined here, and will soon depart (behind 4096 and 8470) for Vancouver as train number 5.
John Freyseng

Canadian Pacific heavyweight diner *Abergeldie* is iced as crews service and combine the Toronto and Montreal sections of the Vancouver-bound *Expo Limited* at Sudbury, Ontario, on May 26, 1967. The short-lived *Expo Limited* – operated only between the spring and fall of 1967 – was the last stand for CPR's heavyweight sleeping and dining cars, as well as many of the streamlined *Grove* series sleepers.
John Freyseng

Freshly scrubbed in the finest "boat train" tradition, CP
#303, with RS10 8481 and a pristine tuscan consist
makes its shipside connection with CP Great Lakes
Steamship *Assiniboia* at Port McNicoll, Ontario, on
September 12, 1964. Operating twice-weekly from
Toronto, #303 connected with the Wednesday and
Saturday Port McNicoll – Fort William sailings of
Canadian Pacific Great Lakes Steamships *Assiniboia*
and *Keewatin.* Return sailings arrived at Port McNicoll
Monday and Thursday, and connected with Toronto-
bound boat train #304. CP's Great Lakes Steamships
passenger service and the connecting boat trains were
discontinued in 1965.
John Freyseng

Handsomely dressed in olive-green and black, with
freshly striped tires and handrails neatly trimmed
in white, high-stepping CNR K-5-a Hudson 5703
makes a spirited exit from Toronto Union with
train #10, the all-stops local to Belleville, Ontario.
Number 10 is a far cry from the high-speed, premier
Toronto-Montreal passenger trains that 5703 and
her four sisters were built to handle, but the aging
Hudson handles the assignment with the same
authority. Originally condemned to the torch,
5703 was reprieved when scrappers mistakenly cut
into sister 5700, officially earmarked for preservation. Masquerading as 5700, the renumbered 5703
was displayed for many years at the National
Museum of Science & Technology in Ottawa, but
has since been moved to St. Thomas, Ontario. Sister
5702 is displayed at the Canadian Railway Museum
in Delson, Quebec.
Robert J. Sandusky.

The Toronto skyline shimmers in the distance as VIA LRC 6900 and five Tempo cars hurry through the Exhibition GO station with Windsor-bound #75 on October 18, 1984. *Greg McDonnell*

With a heavy complement of mail and express on the head end, CN U-2-g Northern 6213 soars across Sixteen Mile Creek in Oakville, Ontario, with Toronto-Niagara Falls #101 on October 17, 1958. One of a half-dozen CNR Northerns preserved, 6213 is displayed in Exhibition Park in Toronto. *Paterson-George Collection*

While a blizzard rages outside, Ontario Northland FP7 1518 rests quietly in the warmth of the diesel shop in North Bay, Ontario, on the night of January 11, 1975. Under the protection of the traditional blue flag, diesel maintainers tend to the 1518's mechanical needs, and come morning, the aging FP7 will be ready for the road.
Greg McDonnell

Beyond the Blue Flag ... Men at Work

"A blue signal displayed at one or both ends of an engine, car or train, indicates that workmen are under or about it, when thus protected it must not be coupled to or moved ... " Traditionally, the requirements of Rule 26 of the Uniform Code of Operating Rules have been fulfilled by a blue flag by day and a blue light by night. Behind those blue signals toil many of railroading's unsung heroes.

At division points and terminals, in yards and shops, rip tracks and roundhouses, rain or snow, night or day, whether it's 40 below or 95 above, the work goes on. It's hard, dirty work that receives precious little recognition and no glory. Inside the shops and roundhouses, the air is heavy with the smell of oil, grease, smoke and sweat, and filled with the cacophonic din of an ongoing heavy metal symphony. Power tools scream, engines clatter and whine, bells clang, machinery drones, compressors pound and the ear-piercing ring of steel on steel keeps time.

For those at work outdoors, blue flags offer protection from unexpected moves, but Rule 26 provides no shelter from the elements. Regardless of the conditions, carmen and labourers, maintainers, hostlers, shopmen, machinists and electricians carry on, inspecting cars, trains and locomotives, fuelling and watering, effecting on-the-spot repairs, performing brake tests and keeping the railroad fluid – 24 hours a day, 365 days a year.

Painters sand the blistered nose of CP FP7 4074, berthed in Ogden Shops in Calgary, Alberta for a class-one overhaul on June 19, 1974. While the work on the road-weary F-unit is extensive, workers in the adjacent bay are performing a near-miracle, restoring fire-damaged CPA16-4 4053 to operating condition.
Greg McDonnell

"From PSC to Toronto Merry Xmas" reads the message chalked on a rebuilt traction motor and wheel set resting on the shop floor of CN's Spadina roundhouse in Toronto on December 27, 1980. Three days after Christmas, shop crews at Spadina have yet to unwrap the several-hundred pound Christmas present delivered from their co-workers at Pointe St. Charles shops in Montreal.
Greg McDonnell

On the eve of her longest night, CPR FA2 4087 rests inside the diesel shop at St. Luc, Quebec, on May 27, 1975. Positioned on the drop table for truck work, the aged cab awaits repairs that will never come. Workmen have chalked repair instructions on her trucks, but instead of authorizing the work, higher powers have signed the 4087's death warrant. The repairs will not be performed, and in the morning, CP 4087 will be shoved out onto the lengthening scraplines outside.
Greg McDonnell

Just in from a break-in run on the CP Rail Toronto
Transfer (in the company of RSD17 8921), freshly
shopped CPR G5a 1201 rides the turntable at
CPR's John Street roundhouse in Toronto, Ontario,
on April 26, 1976. The last engine built by CP's
Angus Shops, 1201 was overhauled at John Street
by Ontario Rail Association members in prepara-
tion for a new career working passenger excursions
out of the national capital.
Greg McDonnell

Framed in the round-house doorway, CPR N2b Consolidation 3722 rides the turntable at Port McNicoll, Ontario on March 12, 1960. By the winter of 1960, the small Georgian Bay terminal had become a mecca for steam fans, as the trio of 2-8-0's stabled at Port McNicoll – N2's 3632, 3722 and spare engine, M4d 3422 – were the last regularly assigned CPR steam locomotives in Ontario. On April 30, 1960, 3722 handled the Port McNicoll-Orillia wayfreight for the last time, closing the book on regularly scheduled operation of CPR steam in Ontario.
Robert J. Sandusky

Resplendent in a fresh coat of olive-green and gold, CNR RS3 3005 shares a quiet moment with sisters 3003 and 3000 in the Mimico, Ontario, round-house on October 30, 1960. One of the last units to be painted in the traditional green and gold, 3005 would never wear the "new-image" black and vermilion paint, and would later gain notoriety as the last RS3 "in the old."
Don McQueen

With their distinctive noses lined in a row, VIA
LRC's 6909, 6903 and 6900 await attention in
adjoining stalls of CN's Spadina roundhouse in
Toronto, Ontario, on May 21, 1985.
Greg McDonnell

VIA FP9A 6508 has its windshield washed
as diesel maintainers and car-knockers
descend upon #4 at Edmonton, Alberta, on
May 1, 1980. During the scheduled 20-minute
stop at Edmonton, the eastbound *Super
Continental* will be inspected, fuelled, watered
and recrewed before resuming its transcontinental journey.
Steve Bradley

The blue light of the shop crew's kerosene lantern glows in the snow below the cab window and streaks of lantern light mark the movements of diesel maintainers and carmen as the westbound *Canadian* pauses at the CPR station in Winnipeg, Manitoba on February 15, 1976. The brake test is complete, and once FP9 1411 (notable for its centennial-year duty as engine 1867 on the Confederation Train that toured Canada), boiler-equipped GP9 8515 and trailing FP9 1405 are fuelled and watered, #1 will get the highball to roll westward.
Greg McDonnell

With an experienced hand on the independent
brake, the hostler at CN's Halifax, Nova Scotia,
roundhouse gently eases FPA4 6782 up to the sand
pipe on May 1, 1975.
Greg McDonnell

Repeating a timeless ritual, a CNR engineer uses a handful of cotton waste to wipe the headlight and class lamps of Northern 6177 before departing Levis, Quebec, with train #1, The Maritime Express on August 12, 1956. *John Welsh (Carleton Smith Collection)*

Refugees from dieselization in the U.S., a number
of American 0-8-0's and Mikados immigrated to
the coal fields of Cape Breton in the closing years
of the steam season. Eight-coupled power from
such roads as Chicago & Illinois Midland, Detroit
& Toledo Shore Line, Wabash, New York Central
and Pittsburg & Lake Erie found new life in the
employ of Nova Scotia coal roads Cumberland
Railway & Coal, Old Sydney Collieries and Sydney
& Louisburg. A long way from the smoky environs
of Lang Yard in Toledo, Ohio, Sydney & Louisburg
0-8-0 86, formerly Detroit Toledo & South Shore
110, slakes her thirst at the Glace Bay water tank
on May 19, 1960.
James A. Brown

Rods down, CNR S-2-c Mikado 3587 and her crew strike a traditional pose beside the coal tower at Nutana Yard in Saskatoon on August 16, 1959. As evidenced by the oil bunker in the tender, 3587 has no need for the coal tower. Unfortunately, CN will soon have no need for the 3587, and like the row of old locomotives in the background, the MLW-built Mike will be stored, never to run again.
Jim Walder (John Riddell Collection)

On a cold, snowy night in March 1971, CN GP9
4222 and SD40 5133 idle beneath the sand tower
at Port Mann, British Columbia. The dark forms
of two rusting water towers stand ghostlike against
the winter sky, stark reminders of the steam bani-
shed from this terminal by the likes of 4222 – and
sister 4353 further down the line – little more
than a decade earlier.
Larry Russell

Trudging through ever-deepening snow, the carman
at Hawk Junction, Ontario, adds oil to the journals
of Algoma Central flanger 10120, tied to the tail
end of Hearst-Sault Ste. Marie passenger train #2
on January 19, 1965.
James A. Brown

Carefully laid out on rags and cardboard, pistons, power assemblies and other Alco 251 parts clutter the area as workmen in the BC Rail shop at Squamish, British Columbia, perform remedial surgery on BCR RS18 622 on January 29, 1990.
Greg McDonnell

On November 26, 1963, the CNR Stratford Shops' venerable 200-ton Morgan overhead crane cradled a freshly shopped locomotive one last time as 6218 rode high above the shop floor. The last locomotive to be overhauled at Stratford, CNR 6218 replaced 6167 in fan trip service in 1964, while the Stratford Shops were closed in April of the same year and sold to Cooper-Bessemer. Stratford's final graduate performed excursion duties throughout the east until her retirement in 1971, and is now displayed in Fort Erie, Ontario.
James A. Brown

CP FP7 4071 inches carefully out of stall 29 of the roundhouse at Cote St. Luc, Quebec, while the hostler prepares to move 4066 off the turntable on August 7, 1970. Along with E8 1802, the two FP7's will soon depart on "The Herder," a power move to the Glen Yard in Westmount, while Train Master 8917 continues to slumber in the darkness of stall 26.
Greg McDonnell

Weeds grow high above the turntable leads and the big wooden doors creak hauntingly in the wind; the machines are stilled, the stalls are empty, and CN's Spadina roundhouse is closed for good. Still faithfully gripping a rusting rail, the altered message on a faded blue flag says it all: "NO MEN AT WORK."
Greg McDonnell

Winter

Spiked across a vast and diverse land, Canadian railways are pitted against innumerable challenges, but it is nature that subjects the nation's railroads, and railroaders, to the supreme test. Between November and April, nature plays her trump cards, temperatures plummet, the skies darken, and winter unleashes all of its fury upon the land. Snow buries mountain passes, blows across the prairies in blinding blizzards, piles up in drifts, fills in cuts and freezes hard. In the face of nature's wrath, the trains continue to run, slogging through drawbar-deep snow, exploding through drifts, and forging through raging snowstorms when all other forms of transportation have succumbed.

Snow chokes terminals, clogging switches, neutralizing CTC signals and plugging yards. While harried train dispatchers attempt to work miracles, sectionmen armed with shovels and switch brooms brave the elements to restore order the only way possible – with cold, hard work.

Out on the line, snowplows – shoved by two, three or even four units at "maximum safe speed" – are dispatched to do battle with deepening drifts. Perched in the cupolas of ancient wedge plows that ride like bucking broncos, plow foremen work flanger blade and wing controls, shout operating instructions over the radio to the engineer (whose vision is totally obscured by flying snow), and pray that the plow will break through the next drift without flipping, derailing, or taking to the field.

And then there is the cold – bone-chilling, mind-numbing cold; cold that freezes trainlines and fractures 132-pound rail like brittle toffee; cold that registers in double digits below zero, and in wind-chill factors that prompt weather warnings that "exposed skin will freeze in less than five minutes"; cold that makes even simple jobs, like replacing hose-bag gaskets or performing brake tests, almost painful, and lengthy jobs, like changing out broken rails, down right dangerous.

Winter tests the mettle of both man and machine, and the dramatic contest results in a gruelling season of spectacular railroading.

Illuminated by the setting sun, a heavy ice-fog
casts an eerie orange glow as CN Work Extra 4528
Snowplow plows south of Centralia, Ontario, at
16:20 on December 11, 1979.
Greg McDonnell

Driftbusting they call it, and CN #581 is doing just that as M420 3500, F7Bu 9195 and GP9 4381 slog into heavy drifting at "Mustard Cut," just north of Kippen, Ontario, at 15:09 on March 3, 1987.
Greg McDonnell

Living up to its infamous reputation, the snow in Mustard Cut literally swallows CN M420 3500 as #581 struggles southward. Virtually all of the snow accumulated on the front of the 3500 has been encountered since the train entered onto the Exeter Subdivision at Clinton, Ontario, less than 10 miles back.
Greg McDonnell

East of Harriston, Ontario, CP Work Extra
4223 Snowplow bursts from the trees, plowing
the Teeswater Subdivision on January 26,
1982. Notorious for severe snow conditions
and very heavy drifting, the 67.9-mile
Teeswater Subdivision, running westward
from the junction with the Owen Sound
Subdivision at Fraxa to Teeswater, was
abandoned in 1988.
Greg McDonnell

Speed drops rapidly as CP Plow 400780 churns
through a heavily drifted cut just north of Lakeside,
Ontario, at 15:00 on February 10, 1976. The howl
of a GM 16-567B is near deafening as CP FP7 4061
struggles single-handedly to push the plow through
the snow, nearly prow-deep in places. At the throttle
of the venerable F is engineer Jack Sandercote,
whose adept train-handling skills will get Work
Extra 4061 out of trouble – barely.
Greg McDonnell

Just east of Lucan, Ontario, CN Work Extra 9178 Snowplow is snowbound at mileage 14 on the Forest Subdivision on February 1, 1978. Viewed from the roof of the plow, sectionmen are moving in to shovel out snowbound CN F7Au 9178, GP9 4574 and Jordan plow 55614, while RS18 3744 and the van wait in the distance.
Greg McDonnell

Up ahead, CP RS3 8448 is hopelessly mired, up to its cab windows in snow, but conductor Harold Spence appears unconcerned as he looks south from the rear platform of van 438854, hoping to spy the rescue plow dispatched from Toronto. Running van hop to Orangeville for several days of plow duty on the Bruce branches, Extra 8448 North became stuck in unexpectedly severe drifting just south of Brampton on January 31, 1971. "Spence", a siding on CP's Mactier Subdivision was named in honour of Harold, and brothers Garfield and Roy, all CPR train service employees working out of Toronto.
James A. Brown

Sectionmen shovel out the last few feet as CP Work
Extra 8740 Snowplow comes to the rescue of the
Teeswater Subdivision wayfreight (SW1200RS
8168, two empty stock cars and a wooden van),
snowbound north of Arthur, Ontario, on
March 27, 1970.
James A. Brown

The going is easy as CP RS10 8467, F7B 4427 and
RS3 8446 drift into Durham, Ontario, with a
Walkerton Subdivision plow extra on January 28,
1978. The 37.2-mile Saugeen-Walkerton line was
abandoned in two stages between 1983 and 1984.
Greg McDonnell

Sprinting westward at 50 mph, the
"London Pick-up," CP Extra 5522 West, explodes
through drifted snow at mileage 51 on the Galt Subdivision,
six miles east of Galt, Ontario, at 15:50 on
January 31, 1987.
William D. Miller

Labouring through prow-deep snow, Work Extra
4063 Snowplow crawls northward on the St. Mary's
Subdivision, just out of Embro, Ontario, on
January 31, 1978. Alarm bells are ringing in the
cab of the 4063 as trailing unit C424 4245 has
succumbed to the snow, leaving the aged cab unit
and RS10 8592 to go it alone.
Greg McDonnell

Advancing across the landscape like tidal wave of
snow, CN Work Extra 4528 Snowplow storms
through "Mustard Cut" at sunset on
December 11, 1977.
Greg McDonnell

Pure Prairie Railroading

There is no mystery to the attraction of railroading in the spectacular setting of the mountains, amid the rugged beauty of the Canadian Shield, or even on the densely trafficked mainlines of the East, but prairie railroading is quite another matter. As those who have ventured west of the Lakehead or east of the foothills can readily testify, the lure of prairie railroading is difficult to define . . . and impossible to deny.

The lure is in the magic of prairie branchlines, where railroading is seemingly frozen in time and history is close to the surface. History as tangible as the Canadian Northern "sheaf of wheat" symbol cast in concrete on an old station, or as subtle as roundhouse ruins, or the barely perceptible remains of a long-abandoned branch-line. History that manifests itself in the unmistakable Grand Trunk Pacific architecture of a small town depot, or the discovery of an ancient wooden car dating to the heady days of prairie colonization when, "wise men of the east" went west by CPR.

The lure is in the drama of transcontinental hotshots that thunder across the prairie, hell-bent for the coast, with screaming SD40's trailing piggybacks, containers and auto racks, and in the endless parade of grain trains that drag prairie harvests to Pacific ports and Lakehead elevators. It's in the awe-inspiring beauty of a prairie sunset and the intense loneliness that tugs at the emotions as the total darkness of a moonless night swallows a desolate branchline. From Molson to Morris, Gladstone and Neepawa, to Broadview, Swift Current, Shaunavon and Taber, Lethbridge, Lacombe, Big Valley and Bon Accord, the siren call of prairie railroading – pure railroading in an unspoiled land of extremes – is irresistible.

Labouring out of a spectacular prairie sunset east
of Arden, Manitoba, CP SD40-2 5700 single-
handedly drags an eastbound grain train – running
as #74 – through the big curve at mileage 48 on the
Minnedosa Subdivision at 21:23 on July 22, 1980.
Greg McDonnell

CN "Ten-hundreds" 1046, 1041, 1019 and 1009
tiptoe across the South Saskatchewan River at
Fenton, Saskatchewan with, Hudson Bay-Prince
Albert mixed train M293 on October 16, 1976. CP
Rail trains also used this portion of the Tisdale
Subdivision, exercising trackage rights between
Northway and Prince Albert.
Andrew J. Sutherland

Silhouetted against a golden prairie sunset, CN Extra 1048 East kicks up the dust, working through Bayard, Saskatchewan with the weekly (or as required) Gravelbourg Subdivision grain drag on October 1, 1976. Both ends of the 118.9-mile Gravelbourg Subdivision (including the trackage through Mitchellton and Bayard) have been abandoned, but the 57.3-mile midsection between Mossbank Junction and Tyson has been taken over by CP Rail as part of the overall rationalization of the prairie branchline network. *Andrew J. Sutherland*

A smudge of black exhaust on the horizon marks the head end of CN Extra 1048 East as GMD1's 1048, 1038, and 1022 struggle to move full tonnage over the Gravelbourg Subdivision near Mitchellton, Saskatchewan, on October 1, 1976. Due to weight restrictions on the light rail, the train is predominantly 40-foot grain boxes, but a small cut of lightweight "Trudeau hoppers" can be seen several cars ahead of the caboose. *Andrew J. Sutherland*

The lanky profile of CNR U-1-a 4-8-2 6010 is
accented against the crisp blue prairie sky as the
beetle-browed Mountain works an eastbound extra
through Dacotah, Manitoba on July 30, 1959.
Paterson-George Collection

With F7Au 9152 and the mandatory cut of express
reefers and container flats on the head end, CN
Thompson-Winnipeg train #90 is just out of
Kamsack, Saskatchewan, at 07:50 on September
22, 1976. Maids of all work, the Winnipeg-
Thompson-Churchill passenger trains regularly
carry mail and supplies for remote northern
settlements, in refrigerator and baggage cars,
containers and on piggyback flats.
Andrew J. Sutherland

Bound for the former Spokane International gateway at Kingsgate, British Columbia/ Eastport, Idaho, on January 24, 1982, snow-packed CP Rail SD40-2 5625 and a sister 5600 battle heavy drifting and bitter cold with CP/UP Calgary-Hinkle, Oregon, runthrough #979 at Fort Macleod, Alberta. The CPR has been funnelling traffic through the remote gateway to the U.S. Pacific Northwest since opening of the Spokane International in 1906.
Richard Yaremko

Teamed up with a CP GP9, a SOO Line FP7 and
F7B move eastbound tonnage across the frozen
prairie landscape near Swift Current, Saskatchewan,
in March 1965. Leased to help their parent road
cope with the seasonal grain rush, the SOO units
were a rarity on CP lines in 1965. Twenty years
later, SOO SD40-2's would be regular visitors on
CP-SOO runthrough trains.
Robert J. Sandusky

On October 6, 1979, NAR #40, with GP9's 207, 205 and 204, is in the clear at Morinville for McClennan-bound train #31, behind GP9's 211 and 206, along with SD38-2 402. Morinville, 19.5 miles out of Dunvegan on the Edmonton Subdivision, was a common meeting place for the two freights, which handled tonnage between Dunvegan and McClennan.
Richard Yaremko

It may be sunrise, but NAR steam operations are in twilight as Northern Alberta Railways 2-8-0 72 goes to work in Dunvegan Yards on a crisp September morning in 1960. Built by CLC in October 1927, number 72 is fast approaching her 33rd, and final, birthday. NAR dropped the fires of most surviving steam locomotives on September 30, 1960 (although 2-8-0 74 was used sporadically in yard service during October 1960), and along with all remaining NAR steam engines, number 72 was sold for scrap. Sister engine 73 escaped the torch, as she was purchased from Stelco by the Rocky Mountain Division of the Canadian Railroad Historical Association. The sole surviving NAR steam locomotive, the 73 was restored to operating condition and is now preserved in Edmonton in the care of the Alberta Pioneer Railway Association. *Robert J. Sandusky*

Working the south-bound Barrhead wayfreight on September 16, 1960, NAR 72 switches the UGG grain elevators at Alcomdale, Alberta, 4.8 miles south of the Barrhead Subdivision's junction with the mainline at Busby. The 26.1-mile Busby to Barrhead line was chartered as the Pembina Valley Railway in 1926 – only one year before the Consolidation was outshopped from CLC's Kingston works as Pembina Valley Railway 72. *Robert J. Sandusky*

Charging eastward under a threatening sky,
CPR G3d Pacific 2343 storms through Indian
Head, Saskatchewan, with eastbound grain
boxes on July 21, 1959. Built by MLW in
September 1926, the handsome Pacific was
not scrapped until March 1964, one year
after sister 2341 was acquired by the CRHA
for display at the Canadian Railway Museum
at Delson Quebec.
Paterson-George Collection

Slicing across the golden prairie with a matched A-B-A of tuscan-and-grey F-units and 17 cars, CPR #1, *The Canadian*, breezes through Gull Lake, Saskatchewan, on October 22, 1964. The lead unit, FP7 1418, sports an experimental snow-plow pilot, as well as the distinctive roof-top icicle breakers (to protect dome glass from ice hanging from tunnel ceilings) and rotating searchlight outfitted on all 1400's assigned to transcontinental domeliners.
James A. Brown

Bringing up the markers on CPR #1 at Gull Lake, Saskatchewan, Budd-built sleeper/buffet-lounge/dome-observation car *"Tweedsmuir Park"* glides past the station on October 22, 1964. In 1957, sister car *"Fundy Park"* was destroyed in a rear-end collision at Gull Lake; more than two decades later, a sectionman discovered the name/description plate from inside *"Fundy Park"* buried in the dirt near the wreck site.
James A. Brown

The grain harvest is in full swing as CP Rail GP9's
8698 and 8483 lead a multi-coloured assortment
of grain hoppers through freshly cut fields east of
Plum Coulee, Manitoba, on August 19, 1984.
Greg McDonnell

The drone of normally aspirated 16-645's drifts over the prairie as CN GP38-2's 5526, 5513 and 5519 lumber into East Warman, Saskatchewan, under threatening black skies on July 15, 1987. Extra 5526 West is operating on the Aberdeen Subdivision, a secondary line that was built as part of the Canadian Northern's Winnipeg-Dauphin-Edmonton mainline.
Robert J. Gallagher

Seven miles east of Empress, Alberta, CP GP9
8536 leads the eastbound Empress Subdivision
wayfreight over the South Saskatchewan River
bridge on September 23, 1974. Traffic on the
Empress Subdivision was sporadic at best by this
date, and usually confined to grain and the occa-
sional stock movement. On this day, however,
Extra 8536 East is made up entirely of empty
bulkhead flats used to transport pipe. The Empress-
Leader, Saskatchewan portion of the Empress
Subdivision was abandoned in 1990.
Doug Phillips

The depot dog is on patrol around the back; milk cans, LCL and express packages are piled on the platform; and there are at least two passengers on hand at the old Grand Trunk Pacific station as CNR #66, the Swan River-Regina local, drifts into the Qu'Appelle Valley settlement of Lebret, Saskatchewan, on July 2, 1960. Still wearing the same coat of green and gold paint, GMD1 1063 would gain notoriety 24 years later as the last operating unit in the old colours. On August 20, 1984, an era came to an end as the 1063 was moved to Transcona Shops to trade her faded green dress for a coat of glossy black with red trim. *Robert J. Sandusky*

The moon is still visible overhead as CN GP9's 4307, 4117 and 4108 approach Portage la Prairie, Manitoba, with an eastbound manifest on October 23, 1964. *James A. Brown*

On its first run after a major overhaul at Transcona, CN F7Au 9165 is less than 24 hours out of the shop as #532 grinds to a halt beside the former Northern Pacific station at Morris, Manitoba, at 14:50 on February 1, 1981. Leading GP9's 4322 and 4120, the freshly shopped F-unit is bound for Noyes, Minnesota, with the daily Burlington Northern connection out of Winnipeg. Although the line between Winnipeg and the Minnesota border is now CN's Letellier Subdivision, the Morris station is not the only surviving vestige of Northern Pacific's once considerable presence in Manitoba. Number 532 is stopped at Morris for a meet with Extra BNML 2 North, the daily Burlington Northern Manitoba Limited local between Winnipeg and Noyes. *Greg McDonnell*

Tied up at Belmont, Manitoba, in accordance with orders prohibiting operation on the 60-pound rail of the Hartney Subdivision during the heat of the day, CN GMD1's 1052 and 1070 will soon be back at work. The shadows are growing longer, the heat is easing, and the crew is preparing to resume the journey to Elgin, setting out empty grain boxes on the westward trip and lifting loads on the return.
Greg McDonnell

Working westward on the Wood Mountain
Subdivision, CP GP9's 8657 and 8689 slip quietly
past the church at Lakenheath, Saskatchewan, on
July 25, 1977.
Andrew J. Sutherland

Parked in front of the CPR station at Hussar, Alberta, CP caboose 436675 and an arch-roofed combine wait while M713, the once-a-week Bassano-Calgary mixed train, switches the grain elevators on October 8, 1964.
James A. Brown

Heading west on the Wood Mountain Subdivision on July 25, 1977, CP GP9's 8657 and 8689 lead a sizable train of hoppers and 40-foot grain boxes around the west end of Twelve Mile Lake, on the approach to Flintoft, Saskatchewan. Punctuated by wooden CP van 437254, the Assiniboia-based wayfreight worked the 64.6-mile Ogle-Mankota line about once a week. Visible just above the van, Flintoft generated traffic from three grain elevators, as well as carloads of clay drawn from a local deposit of very pure kaolinite.
Andrew J. Sutherland

While snow squalls rage across the foothills of the
Rockies, the sun breaks from behind the clouds
just long enough to shine on CP GP9 8661, H16-44
8710 and a short train eastbound at Cowley,
Alberta at 17:30 on March 13, 1974.
John Sutherland

While Spokane International 2-8-0's and Mikes were turned back at Yahk, and in later years, SI RS1's ventured no further than Eastport, a 1978 CP/UP runthrough agreement provided for reciprocal power pooling on Calgary-Hinkle, Oregon, trains 979 and 980. In accordance with the agreement, UP SD40-2's 3412 and 3656 team up with CP EMD-built SD40-2 5656 and C424 4242 on Hinkle-Calgary runthrough #980 at Cowley, Alberta, on June 12, 1983. Attesting to its assignment to the CP pool, UP 3412 has been outfitted with CP ditch lights.
Richard Yaremko

The head-end crew is out for the camera as
Belpaire-boilered CNR S-2-c Mikado 3590
storms into Saskatoon with #412 on
July 20, 1959.
Paterson-George Collection

At the height of a late-season prairie blizzard, Manitoba & Saskatchewan Coal's ex-CPR 2-8-0 3522 clumps across the CNR diamond at Bienfait, Saskatchewan, on March 23, 1964, with a cut of cars for the CPR interchange. One of the few surviving revenue steam operations in the land, M&SC kept coal smoke over the Saskatchewan prairie for a few more seasons, alternating the 3522 in service with ex-CPR 0-8-0 6947. The 3522 has been preserved at Bienfait; 6947 is the property of the Alberta Pioneer Railway Association of Edmonton; and M&SC's ex-CP 0-6-0 6166 is displayed at the Western Development Museum in North Battleford, Saskatchewan.
Robert J. Sandusky

Laying down a trail of heavy black oil smoke, CNR Alco-built Consolidation 2128 approaches North Battleford, Saskatchewan, with mixed train M354 on July 18, 1959. Built in 1911, the 2128 was outshopped from Alco's Brooks works as Duluth Winnipeg & Pacific 2128.
Paterson-George Collection

Heat waves shimmer off the prairie as CP GP38 3009 and GP9 8697 climb over the horizon and approach the abandoned elevator at Bures, Saskatchewan, with the westbound "Weyburn Tramp" on the Amulet Subdivision on June 9, 1977. The aptly named Tramp ran all directions out of Weyburn on various days of the week, and on this occasion was making a Cardross Turn. The Cardross-Crane Valley portion of the Amulet Subdivision was abandoned on July 29, 1982.
John Sutherland

Running cab hop in a spring snowstorm, CP GP38 3014 leads the "Wishart Turn" through West Bend, Saskatchewan, on April 18, 1981. Operating out of Wynyard, the weekly "Wishart Turn" generally ran Tuesdays, but this is a rare weekend run, being made to lift 20 loads of grain from on-line elevators. Train crews despised these turns, on account of low mileage and the slow, 15 mph track speed over the length of the subdivision. Built by the Northwestern Railway Company between 1927 and 1929, the 26.9-mile Wishart Subdivision was officially abandoned on December 31, 1982.
Andrew J. Sutherland

At the end of the line, the "Weyburn Tramp" switches the elevator at Cardross, Saskatchewan, on June 9, 1977. Coupled to modernized wooden caboose 437161, CP GP38 3009 and GP9 8697 have already run around their train, and upon completion of the work at Cardross, the Tramp will begin working its way back to Weyburn.
Andrew J. Sutherland

Somewhat worse for wear, flat-faced CPR G3h
Pacific 2451 pulls an eastbound drag into
Chater, Manitoba, on July 21, 1959. Steam is
making its final stand on the prairies and the
road-weary Pacific's working days are almost
at an end.
Dick George (Paterson-George Collection)

On the last day of May 1974, CP Extra 4104 West storms through Cowley, Alberta with back-to-back CPA16-4's 4104 and 4105 handling overflow traffic out of Lethbridge. On the train sheets and consist reports, the train, made up of lumber empties and Slave Lake ore, is just a westbound drag, but in the annals of diesel history, it will go down as what is believed to be the last A-A set of C-lines to run anywhere. Privately-owned, the 4104 has been preserved in operating condition at High River, Alberta.
Andrew J. Sutherland

Returning from Lac du Bonnet, Manitoba with
two loads of cut granite and an empty grain box,
CP GP9 8648 pulls around the wye at Molson,
Manitoba, on August 21, 1984. While the wayfreight performs a few switching chores, the
Molson operator is copying train orders for its
return to Winnipeg on the Keewatin Subdivision
mainline as Extra 8648 East.
Greg McDonnell

Running as Second 840, CN GP40-2L's 9486, 9563
and GP9's 4300 and 4228 work a westbound drag
through Golden Stream, Manitoba, on July 23, 1980.
Although most of the train consists of grain
empties, the first five cars are heading west for
hay-loading as prairie farmers assist their Ontario
counterparts through a serious drought and
resulting feed shortage.
Greg McDonnell

The whine of dynamic brakes fills the Peace River
Valley on June 30, 1979 as NAR Extra 402 North
descends the 2.4 percent grade at Peace River with
tonnage bound for CN's Great Slave Lake Railway
at Roma Junction. While most of the northbound
cars are consigned to GSLR points between Roma
Junction and Hay River, Northwest Territories, the
drilling mud loaded in covered hoppers on the
head end of the train will be transhipped to a barge
and moved to drilling sites in the MacKenzie River
delta. Purchased specifically for McClennan-Roma
Junction service, NAR SD38-2's 401-404 were the
road's largest power, as well as the only NAR
diesels equipped with dynamic brakes.
Richard Yaremko

Storming out of Edmonton, Northern Alberta Railways 4-6-2 161 has a dozen cars in hand, working train #1, bound for Dawson Creek, British Columbia. The 495-mile run to Dawson Creek will consume almost 18 hours and include 74 scheduled stops. A new road number and NORTHERN ALBERTA stencilled across the tender cannot disguise the ancestry of the NAR's only Pacific. Purchased in 1947, NAR 161 is ex-CPR G2s 2563. *Paterson-George Collection*

Treading carefully through marshes along the shore of Lesser Slave Lake, CN F7Au's 9159, 9169 and F7Bu 9192 roll grain empties westward on the former-NAR Slave Lake Subdivision near Wagner, Alberta, on July 7, 1985. *John Sutherland*

Suffield sunsets... Silhouetted in the setting sun
on June 20, 1973, the CPR station at Suffield,
Alberta stands guard at the junction of the Suffield
Subdivision and the mainline west of Medicine
Hat. Office signal "SU", Suffield handles train
orders for both the mainline Brooks Subdivision
and the Suffield Subdivision, a lightly trafficked
grain branch. For the moment, though, all is quiet,
the order board is clear, and the Suffield operator
maintains a lonely vigil on a peaceful evening just
shy of the summer solstice.
James A. Brown

At the same hour, four years and four days later,
Suffield is not so quiet, as #965 hurtles into the
sunset with multi-level loads of westbound Fords
on June 24, 1974.
Doug Phillips

Bound for the Pacific coast, "Trudeau hoppers" loaded with export grain trace the Bow River as CP Extra 5727 West snakes through Morant's Curve east of Lake Louise, Alberta, on January 24, 1990. Technically mileage 113 on the Laggan Subdivision, the picturesque location has – unofficially at least – been named in honour of the man whose numerous photographs have made this scene familiar to generations of Canadians, famed Canadian Pacific photographer Nicholas Morant.
Greg McDonnell

Mountains

Looking west from the prairie's edge, the brooding, snow-capped peaks of the Rockies loom large in the distance, their breathtaking beauty only hinting at the magnificence – and perils – that lie ahead. Canada was not truly a nation until a railroad was scratched through the Rockies and the successive ranges to the coast. In promoting the need for a railway to the Pacific, John A. Macdonald stated categorically, "Until this grand work is completed, our Dominion is little more than a geographical expression." Indeed, Canada owes an incalculable debt to the thousands of anonymous labourers that blasted rock and notched ledges into sheer canyon walls, built bridges, bored out tunnels and laid down track, driving the railroad through some of the most forbidding territory on the continent.

At Craigellachie, British Columbia, a simple stone cairn hard by the CPR mainline pays subtle tribute to the herculean labour and the lives lost. Its brief inscription reads: "Here was driven THE LAST SPIKE completing Canadian Pacific Railway from OCEAN TO OCEAN November 7, 1885." As the last spike was driven home at Craigellachie, completion of the CPR secured Confederation and established a physical union of the provinces, but in the mountains, the adventure was just beginning.

On the CPR, and on the successive railways built into the mountains – the Grand Trunk Pacific and Canadian Northern, CPR's own Kettle Valley Railway, arch rival James J. Hill's Great Northern, the Pacific Great Eastern, and the remote White Pass & Yukon – railroading is the stuff of legends. Legends of the "Big Hill" east of Field, where the challenging assault on the 4.5 percent grade was exceeded only by the struggle to maintain control on the treacherous descent, and where, under the ever-present threat of runaways, switch tenders guarded "safety tracks" until the 1909 completion of the Spiral Tunnels by-passed "The Hill." Legends of helper districts and pushers and 10-coupled engines double and triple-headed to move tonnage over mountain grades. Legends of trains clinging precariously to ledges carved into the solid rock face of vertical canyon walls. Legends of avalanches, rock slides and wrecks, and of rotary snowplows clearing mountain passes.

Technological advances and engineering feats – such as CP's $500-million Rogers Pass project completed in 1988 – have tamed but failed to conquer the mountains. In places with names that evoke images of the glorious past, the high drama of mountain railroading is played out still, at Revelstoke and Field, Crowsnest and McGillvary Loop, in the Yellowhead and on the former Grand Trunk Pacific at Tete Jaune, Tintagel, Telkwa and Kitwanga. In Kicking Horse Pass and the Illecillewaet River Canyon, mountains that once echoed the exhaust of doubleheaded Decapods reverberate the turbo-charged drone of multipled SD40-2's, while Alco exhaust hangs heavy over BC Rail's "north end" and full-cowl GE Dash 8-40C's labour on the 2.2 percent grades through Garibaldi, Tisdall, Birken and Glenfraser. Based at Pemberton – the last manned helper station in the land – crews aboard PTC-equipped BCR SD40-2's assist southbound freights from D'Arcy to Mons, and throughout the mountains, remote-controlled, mid-train helpers are the order of the day.

In the canyons of the Thompson and Fraser rivers, the parallel CN and CP mainlines (on opposite banks for most of the distance between Savona and Hope) play host to some of the most treacherous railroading in all Canada. Even the names are foreboding: Black Canyon, China Bar, Hell's Gate and Deadman's Creek. With regularity, the mountains flex their muscles, unleashing tons of rock and mud that briefly reclaim the railway, and on occasion take the lives of railroaders. For the crews aboard the unit trains, transcontinental symbol freights and passenger trains that coil through the canyons, slide fences and rock sheds serve as constant reminders that the mountains may be tamed, but they will not be conquered.

Kicking Horse Canyon echoes the throbbing exhaust of 16-645's as CP Extra 5747 East emerges from a tunnel – hard by the Kicking Horse River – east of Golden, British Columbia, on June 13, 1987.
John Sutherland

Skirting the rushing waters of the Kicking Horse River, CP FP7 4040 and GP9's 8631, 8695 and 8811 labour upstream with eastbound tonnage at Glenogle, British Columbia, on October 12, 1964. Far from Kicking Horse Pass, the 4040 now hauls Montreal commuters as STCUM 1306, while the three GP9's have been chop-nosed and upgraded to 645-powered GP9u's. The 8811 and 8631 have been rebuilt for roadswitcher service as CP 8208 and 8210, while 8695 is now in yard and transfer service as CP 1616.
James A. Brown

Idling on the wye at Rogers, British Columbia, on
February 28, 1987, Beaver Hill pushers 5940, 5570,
5620, 5651, 5652 and 5719 are in position, waiting to
assist a 15,000-ton westbound coal train up the 2.2 percent
grade to Stoney Creek. With the completion of the $500-million
Rogers Pass project in December 1988, westbounds were
diverted to the new 21-mile "Macdonald Track" between Fraine
and Ross Peak (via the Mount Shaughnessy and Mount Macdonald
tunnels), reducing the ruling grade to one percent and allowing
elimination of CP's last manned pusher district. Eastbounds
continue to use the old line – via the Connaught Tunnel and
Stoney Creek bridge – now known as the "Connaught Track."
Steve Bradley

West of Beavermouth, British Columbia, on
September 21, 1990, CP Extra 5929 West
traces the shoreline of Kinbasket Lake on
trackage constructed in 1974 to by-pass
flooding caused by construction of the Mica
Power Dam on the Columbia River.
James A. Brown

High green! CP Plow Extra 5745 East is in the clear
at Donald, British Columbia, and a clear signal
beckons the westbound freight from which this
photo was taken. Still equipped with a "golden
glow" steam-era headlight in January 1978, the
plow is also outfitted with bars on the cupola windows as protection from rocks, and it lacks the
distinctive prow found on most plows assigned to
eastern lines.
Philip Mason

Following the Thompson River, CN F7A 9028 leads SD40 5159 on an eastbound near Martel, British Columbia, in June 1970. Mountain Region crews preferred to have an F-unit in the lead at the time, as the cab units were considered safer in an encounter with a rock slide, an ever-present danger in the treacherous canyons of the Thompson and Fraser rivers. After experience proved that the heavier SD's could plow through many of the smaller slides and generally kept their trucks if derailed, the cab units lost their status as preferred leaders.
Peter Cox

Just west of Boston Bar, British Columbia, in mid-summer 1970, CN SD40's 5140 and 5127 lead #303 over the Anderson Creek trestle. Several months later, 5140 met with disaster not far from this location. On February 15, 1971, 5140 and F7A 9092 collided with a massive rock slide and plunged more than 200 feet into the Fraser River Canyon, along with several cars and three crewmen. The crew members were never found and the wreckage of the 5140 and 9092 remains at the canyon bottom – a grim monument to the dangers of mountain railroading.
Peter Cox

Perched on the boxcar roof, the brakeman on CP Nelson-Cranbrook train #90 watches intently as CP 55074 splashes over submerged rails onto the barge slip at Procter, British Columbia on June 25, 1974. Once loaded aboard the barge, the tug *Melinda Jane* will sail #90's "set-off" northward, up Kootenay Lake to Kaslo and Lardeau. Meanwhile, #90, led by H16-44's 8724, 8555 and leased PNC GP9 135, will swing south and follow the west shore of Kootenay Lake for 35 miles to Kootenay Landing.
Greg McDonnell

Easy does it ... CP H16-44 8726 gingerly moves onto the slip with wooden van 437249 while loading the weekly Nelson-Nakusp wayfreight aboard the barge at Slocan City, British Columbia, on June 7, 1974. Once the entire train (including the 8726) is on board the barge, the tug *Iris G* (seen waiting patiently beside the dock) will ferry the local 20 miles up Slocan Lake to Rosebery. Upon disembarking at Rosebery, the train will continue to Nakusp on the isolated 27-mile Kaslo Subdivision. The last vestige of Canadian Pacific's once expansive marine-rail network serving the British Columbia interior, the Kaslo Subdivision and its Slocan barge connection were abandoned December 31, 1988.
Andrew Sutherland

Approaching the slip at Rosebery, British Columbia,
with CP GP9 8820, a gondola, mechanical reefer
and caboose aboard, the Slocan Lake car float and
tug *Iris G* sail past the public beach on July 27, 1982.
The following day, a crew will taxi from
Nelson, unload the barge and take the train up the
Kaslo Subdivision to Naskusp, 27 miles north and
one valley west.
Philip Mason

Coiling through the S-curve at
Gibbs, British Columbia, with a
heavy train of forest products,
BC Rail Dash 8-40CM's 4602
and 4605, with "remote" 4606
cut in mid-train, drop down the
2.2 percent grade south of
Glenfraser on October 10, 1990.
Brian A. Elchlepp

High above the Fraser River at Glenfraser, British Columbia, BC Rail M630's 722 and 709 – trailing hopper loads of brilliant yellow sulphur – descend the winding 2.2 percent toward Lillooet at 13:20 on August 31, 1986.
John Sutherland

Completing the lowest Rocky Mountain crossing made by any transcontinental railroad in North America, CN SD40's 5232 and 5154 cast a striking reflection working eastbound through Yellowhead Pass, just east of Yellowhead, Alberta, on October 12, 1975. During the 1980s, this portion of the Albreda Subdivision was double-tracked between Jasper West and Red Pass, British Columbia.
Doug Phillips

The turbo-charged howl of GM 16-645's shatters
the calm at Albreda, British Columbia, on
October 16, 1985, as CN SD50F's 5423 and 5429 burst
out of the shadows with westbound potash train #759.
Assignment of just two SD's to the potash unit
train – filled out to full tonnage by a cut of grain
loads added on the head end – is indicative of CN's
easier route through the mountains. Indeed, the
cowled freighters will handle the train through to
the coast unassisted.
Robert J. Gallagher

Dynamic brakes whine and the blue haze of
brakeshoe smoke rises from trailing cars as
CP SD40's 5502, 5512, M630 4514 and SD40
5542 drop downgrade through Cathedral,
British Columbia, on September 20, 1970.
James A. Brown

Trailing nine mail and express cars and a
single 2200-series coach, CP FP9 1407 and
GP9's 8515 and 8510 lead eastbound train #8
across the Stoney Creek bridge on April 17, 1965.
Peter Cox

Just about to duck into Clapperton Tunnel, CP Extra 3005 East, the "Kettle Valley Through Freight" (or KVTF for short) coils through the Nicola River Canyon with a lengthy train of empty wood-chip gons and lumber cars in June 1980. The name is a holdover from better days, though, for the Kettle Valley Railway has been severed east of Penticton and the grandly titled KVTF is but a tri-weekly local working between Spences Bridge and Penticton. In its declining days, the west end of the Kettle Valley Railway – officially known as the 174-mile CP Princeton Subdivision – was supported solely by lumber traffic, and the eastbound KVTF is busy peddling empties to mills at Merritt, Princeton and Okanagan Falls. Not long after this photo was taken, the wood chip traffic was lost to trucks, and in 1989, CP persuaded on-line mills to ship all their traffic by truck. The line was closed in May 1989 and abandoned in July 1990.
Philip Mason

The booming, opposed-piston chant of Fairbanks-Morse diesels fills the air as CP Extra 4065 East, the Midway Turn emerges from the snowshed at Paulson Gap, British Columbia, in June 1972. Boulders and debris piled on the rooftop and strewn down the hillside give graphic testimony to the need for a snowshed at this location. Built in August 1951 as Canadian Locomotive Company 7006, half of the CLC *City of Kingston* CPA16-4 demonstrator set, the FM-design diesel toured nationwide with sister demonstrator 7005 during the fall of 1951. The barnstorming C-lines were sold to CP in December 1951 and renumbered 4064 and 4065 in September 1952. The 4064 was scrapped at Ogden in 1972, after suffering severe fire damage in 1969, but 4065 soldiered on until retirement in 1975 and was saved as part of CP Rail's historic diesel collection.
Stan J. Smaill

An early snow clings to the trees near Moyie, British Columbia, as CP GP9 8805 and H16-44 8554 follow the shoreline of Moyie Lake with an eastbound freight headed for Cranbrook on October 10, 1972.
John Sutherland

Leaking steam envelops CP S2a 2-10-2 5811
as the lanky Decapod clanks past a begrimed
wooden caboose at Field, British Columbia,
in the early 1950s.
John Welsh (Carleton Smith Collection)

Shopmen examine CPR T1b 5925 as the streamlined Selkirk simmers in front of the station at Field, British Columbia, and the westbound *Dominion* takes on mail and changes crews during its 20-minute stop at the Rocky Mountain division point.
John Welsh (Carleton Smith Collection)

White exhaust shrouds the semaphore signals west of the station as CPR T1c Selkirk 5925 makes an impressive exit, departing Field, British Columbia, with the westbound *Dominion*.
John Welsh (Carleton Smith Collection)

Forty-five minutes before midnight, in the land of the midnight sun, White Pass & Yukon DL535E's 104, 106, 101, 107 and 110 hustle two large front-end loaders and a train of concentrate hoppers north of Fraser, British Columbia, at 23:15 on June 9, 1980. After losing the concentrate traffic to trucks, the White Pass shut down in 1982 and remained dormant until 1988. Since then, the WP&Y has operated summer-season tourist trains between Skagway, Alaska, and the border, but the Canadian section of the narrow-gauge line is unused.
Andrew J. Sutherland

Passengers detrain from Pullman-green narrow-gauge coaches as White Pass & Yukon "shovel-nose" GEs 98 and 94 pause at Carcross, Yukon, with the westbound mixed train from Whitehorse to Skagway on June 8, 1980.
Andrew J. Sutherland

Twenty-six years on the island and still going strong, CP Baldwin DRS4-4-1000's 8002 and 8000 lead the "Lake Cowichan Turn" out of Lake Cowichan, British Columbia, with a trainload of logs from the local mill on June 11, 1974. Built by Baldwin in 1948, the 13 DRS4-4-1000's dieselized the Vancouver Island operations of CPR subsidiary Esquimalt & Naniamo and ruled the island until their retirement in 1975. While lead unit 8002 has been scrapped, the 8000 has been preserved as part of CP Rail's collection of historic diesels. *Andrew J. Sutherland*

In weather as dismal as the surroundings, CN SD40-2's 5246, 5252 and 5311 load export coal at the Cardinal River Coal Company's mine near Luscar, Alberta, on July 4, 1987. On the eastern edge of the Rockies, the Cardinal River mine is accessed via a 4.5-mile spur running northward from the Mountain Park Subdivision at Leyland, Alberta. *Doug Phillips*

Amid the lush greens of the Elk River valley near Fernie, British Columbia, "action red" CP Rail SD40-2's 5806 and 5942 roll eastbound coal empties back to the mines in Crowsnest Pass on July 26, 1989.
Steve Bradley

Shouting to the mountains, the deafening cry of Alco 251 and GM 645 engines echoes off distant peaks as CP M630 4570 and SD40-2 5580 crawl into Elko, British Columbia, on June 23, 1974, dragging 15,000 tons of Crowsnest coal. Despite the assistance of mid-train "robot" helpers 5573 and 5614 (radio-controlled by "Locotrol" equipment aboard the 4570), Extra 4570 West is making barely enough speed to flap the white extra flags.
Greg McDonnell

Again, nature has flexed her muscles and taken
out the CN mainline through the mountains. As a
result, CN SD40-2 5283 and SD40 sisters 5214,
5236, 5198 and 5240 are detouring eastbound
tonnage over CP Rail's Thompson Subdivision along
the Thompson River in August 1977. Temporarily
devoid of traffic, the CN Ashcroft Subdivision –
"closed for repairs" – is visible on the opposite bank.
Kenneth R. Goslett

Undaunted by heavy snow that moments earlier caused a light plane to crash-land on a nearby road, BC Rail M630 722 and SD40-2 765 labour up the 2.2 percent grade through Whistler, British Columbia, at 16:11 on January 26, 1990.
Greg McDonnell

Accelerating through the S-curve at Moyie, British Columbia, a quintet of CP CLC's, all " in the print," highball #984 eastward on the Nelson Subdivision in August 1970. Remarkably, while almost all CLC-built diesels have been scrapped, three of the five units in this consist are extant in 1991. Saved as part of the CP Rail historic diesel collection, 4065 was moved to Ottawa in 1990 for display at the National Museum of Science and Technology, while H16-44 8554 is stored awaiting transfer to a suitable home. Not so fortunate, 4471, converted to robot car 1009 in 1972, is believed to be among the hulks languishing in a Calgary scrapyard.
Stan J. Smaill

The sun is just over the mountains and Yahk, British
Columbia, is blanketed with a heavy layer of frost
as westbound CP H16-44's 8722, 8709 and CFB16-4
4451 ease up to a red train-order board just after
dawn on October 21, 1964.
James A. Brown

Living up to their name, and looking every bit like something straight out of the publicity files of Fairbanks-Morse, CP Train Masters 8906 and 8917 swing through the S-curves and sagebrush at Savona, British Columbia, on May 18, 1963, making light work of a westbound drag. After spending much of their early careers based out of Winnipeg, many of the big H24-66's were transferred to general freight service between Calgary and Coquitlam during the early sixties. The desert-like terrain around Savona is still noted for sagebrush, ponderosa pine and rattlesnakes, but the Train Masters are long gone. Preserved at the Canadian Railway Museum in Delson, Quebec, CP H24-66 8905 is the only Train Master left in existence.
Peter Cox

Grinding through Coryell, British Columbia, engineer Walter Holabawick is coaxing every available ounce of power from CPR CPA16-4 4065, a pair of H16-44's and a GP9, as the Midway Turn, "Extra 4065 East," assaults the west slope of Farron Hill in June 1972. The last train has challenged Farron Hill and Coryell is a quieter place, as the Boundary Subdivision was officially abandoned through here on December 9, 1990. *Stan J. Smaill*

Clinging to a narrow ledge notched into the hillside, BC Rail M630's 722 and 709 squeal through a tight curve near Glenfraser, British Columbia, on August 31, 1986. Both of these units were among the 21 BC Rail big Alcos sent to GE as trade-ins on Dash 8-40CM's 4601-4622 delivered in 1990. Their careers were far from over, as GE dispatched most of the units to Mexico, and in 1991, BCR 709 and 722 are hard at work in the employ of Ferrocarriles Nacionales de Mexico.
John Sutherland

Winding along the rocky shoreline of Howe Sound, south of Squamish, British Columbia, PGE RS18's 627, 618 and 604, RS3 576 and ex-Lehigh & Hudson River C420 25 (soon to become PGE/BCR 631) head for North Vancouver with a southbound freight in August 1972.
Kenneth R. Goslett

Overshadowed by a massive extrusion of rock, Second 70, with SD40-2's 5659 and 5587 bracketing M630's 4508 and 4510, works along the shore of a small lake east of Hazell, Alberta on CP Rail's Crowsnest Subdivision on June 21, 1974. *Greg McDonnell*

West of Savona, the CP and CN mainlines parallel one another through the Thompson and Fraser River canyons, encountering some of the most spectacular and forbidding terrain in the process. Meets and "races" between trains of the two roads are a daily occurrence in this territory. In the Thompson River Canyon near Martel, a CPR westbound with a pair of SD40-2's paces a CN manifest headed by a GP38-2 and two SD40's. Facing stiffer grades, but running on faster track, oddsmakers and those familiar with the railroad would likely give the edge to the CPR freight, but few would wager on the chances of staging this particular race, pitting CP 5605 against CN 5605. Heading for the over/under at Cisco on a sunny day in October 1980, CN Extra 5605 West and CP Extra 5605 West are neck and neck.
Doug Phillips

Working westbound grain over one of the most treacherous pieces of railroad in Canada, CN SD50F 5400 and SD40-2's 5256 and 5326 roll through one of the numerous slide zones in the "White Canyon" near Lytton, British Columbia, on August 31, 1986.
John Sutherland

Epilogue

We were hammering eastward, following the CPR Adirondack Subdivision in Stan Smaill's 1954 Chevy, when I first confided the dream of producing a book that would capture the magic and the essence of Canadian railroading. More than 20 years later, this book is the manifestation of that dream, but even more, it is the product of the influence, inspiration and encouragement provided by a number of good friends.

As the old Chevy (affectionately known as "The Heap") rumbled past Adirondack Junction, our discussion set the dream in motion, and Stan's early encouragement provided me with the initial inspiration and confidence to pursue it. In the intervening years, other friends came to share the dream. John and Andrew Sutherland have made instrumental contributions, as have Chuck Begg and my brother, George McDonnell.

On a pleasant night in April 1989, Jim Brown provided the catalyst that finally brought the project to life. An informal gathering at Jim's "Alliston Station" assembled many of the photographers whose work appears here. Long into the night, we savoured vintage Kodachromes of steam-era railroading, while outside, CP Rail SD40-2's hurtled past with transcontinental manifests and northbound drags. It was a memorable experience, blending celluloid images and anecdotes of the glory days with the ear-splitting, earth-shaking drama of contemporary railroading.

At 04:08, we stood on the platform and watched as British Columbia-bound #401 swished past, silhouetted in the bright moonlight. As the van rolled by at track speed, a tiny speck of light from the conductor's lantern waved a highball from the cupola. Jim responded in kind, and from inside the station, the radio crackled, "Highball from the stationmaster, six double-0-seven North." It was never more apparent than at that moment, that the magic that had drawn us trackside generations ago was still alive and well. As the first hints of daylight brightened the eastern horizon, I knew that this book had to be done.

George Roth ushered this book from concept to reality, and as its designer, he has created a strong presentation that preserves the integrity of each image and respects the photographer's original composition. Behind the scenes, many others have assisted with this work: Newt Rossiter, Frank Bunker, Mac Allen, Ron Bowman and Brad Jolliffe provided invaluable information, anecdotes and details, while Jim Lannigan, Tim Sauer and Sheila Stemmler helped with other aspects of this project.

Finally, this book would not have been possible without the encouragement, assistance, understanding, and endurance of my best friend and wife, Maureen. Along with our three sons, Glen, Ted and Andrew, Maureen has endured the long days and late nights, and tolerated the extraordinary demands that this work has placed upon our family life. Maureen has also reviewed text, assisted with the photo selection and provided strong moral support when I needed it most. Friends like that are hard to come by.

The Photographers

Steve Bradley
A railroader at heart, Steve Bradley of Schomberg, Ontario has worked for CN and CP, but currently flies as a Boeing 727 captain with Air Canada. Steve has been photographing trains since the 1960s and now produces railway videos under the name of Rail Innovations.

James A. Brown
Born and raised in Toronto, Ontario, James A. Brown recounts his earliest memory of trains was a CPR steam doubleheader that he witnessed as a terrified three year old accompanied by his father and grandfather. Since then, he has maintained an interest, as well as a career, in railroading. After working summers and holidays as a spare leverman throughout the CPR "Ontario District," Jim graduated from University of Toronto as a mechanical engineer and worked for CN until 1972, when he joined GO Transit as a Rail Operations Engineer. Jim is now the Executive Director of Operations at GO Transit. An active photographer for more than 35 years, Jim's earliest works date to 1954, but he began photography "in earnest" in 1956, and the James A. Brown credit line has appeared below hundreds of photographs in dozens of books and other publications. Founding director of the Ontario Rail Association, Jim negotiated the use of the ORA's ex-CP 4-4-0 136 and other equipment in the CBC production of The National Dream. In 1988, Jim acquired, moved (to a railside site near Tottenham, Ontario) and restored the former Canadian Pacific "Type 2" station from Alliston, Ontario. Jim now calls "Alliston Station" home, and "would need a great deal of convincing to leave!"

Doug Conrad
Doug Conrad has been interested in railroading since childhood. A resident of Halifax, Doug's main interests are in the railroads of New England and eastern Canada, with special affection for Nova Scotia's Dominion Atlantic Railway. Doug has been taking railway photos since the early seventies and has concentrated on activity in and around the Maritime Provinces.

Peter Cox
Peter Cox began photographing railways in 1953, concentrating on Canada and specializing in the West. He has been a serious railway historian throughout this period and is well acquainted with others in this fascinating field. Peter is currently redeveloping an earlier interest in street railways and is active with the trolley cars at Fort Edmonton Park.

Brian A. Elchlepp
A long-time admirer of the locomotive products of Alco/MLW, Brian Elchlepp was drawn to the British Columbia Railway in the early eighties in search of RS18's, hand-me-down C425's, 6-motor Centuries and M-lines. Once north of the 49th parallel, he was captivated by the grace, power and charm of Canada's railways. He has since travelled the country from coast to coast to observe and photograph contemporary Canadian railroading and keeps in close touch with the BCR Alcos.

John Freyseng
A lifelong Torontonian, John Freyseng was born in the city in 1937 and has remained there, but for travel to photograph railways and conduct business. John cites all aspects of railroading, including their history, as a love of his, "second only to the love for his wife, Sandy (a flight attendant with Air Canada for close to 20 years – no love of trains there!) and his four children (who have all been successfully brainwashed)." Numerous business trips to Western Europe and the U.S. have provided John with opportunities to photograph many operations in these countries, complementing his coast-to-coast coverage of Canadian railways. Noted for his exceptional footage of revenue-service Canadian steam, John began shooting movies in 1955 and graduated to 35mm colour slides in 1962.

Bob Gallagher
Bob Gallagher has been engaged in railroad photography for more than 15 years, and is well known for his sensitive and spectacular treatment of prairie railroading. Bob pays special attention to location and scene in his ongoing efforts to capture railroading and its relationship with the land. It's often hard work, and the challenges are great, but the effort is very rewarding.

Dick George and Al Paterson
The names Dick George and Al Paterson are synonymous with Canadian steam. The renowned Paterson-George Collection is widely recognized as the most comprehensive photographic record of Canadian steam locomotives anywhere. Although Dick and Al's cameras were "stored serviceable" with the end of steam, they continued to expand their joint collection of negatives through purchases and trading. While the collection is technically private, its domain has been largely public, as Dick George and Al Paterson have supplied literally thousands of prints to collectors, railroaders, historians, modellers and authors. Black & white negatives dominate the files, but as evidenced here, there is also a good store of colour material within the Paterson-George Collection.

Kenneth R. Goslett
Ken Goslett was born in 1950 in Montreal, a city that boasts not only the headquarters of Canada's two major railways, but the greatest diversity of motive power and operations in the nation. A former CPR employee, Ken has studied and photographed railway architecture in the province of Quebec for both the federal and provincial governments. Ken maintains an active interest in railroads and is a noted authority on diesels and rolling stock of Canadian railways. A long-time volunteer and frequent director of the Canadian Railway Museum, Ken lives in Montreal with his wife and two daughters.

L. Norman Herbert
Norm Herbert has been involved in railroad photography since the early 1950s and has a keen interest in all aspects of railroading, including steam, diesel, streetcars and interurbans. A resident of the Detroit area, Norm has close connections with Canadian railways and has travelled extensively throughout the East, as well as other regions of the country.

Robert E. Lambrecht
Bob Lambrecht has been interested in railroading since 1975. Originally from the Chicago area, he guided "Big R's Chicago Railfan Tours" during his high-school years and later worked as a brakeman-conductor for the C&NW Railroad. Following a service department position with Electro-Motive, Bob moved to Erie, Pennsylvania, where he is employed by General Electric as a project engineer responsible for the Super 7 locomotive line. Prior to their retirement in 1989, Bob focused his attention on photographing the last days of the VIA FPA4's.

Philip Mason
Born in Leicester, England, in January 1950, Philip Mason immigrated to Montreal in 1960. Phil began shooting 35mm colour slides in 1966 and was active in the Montreal railfan community until he moved west in 1975. Employed as a locomotive engineer for CP Rail at Revelstoke, British Columbia, Phil remains an active photographer and observer of railways in western Canada and the United States.

Don McQueen
A noted railway historian, Don McQueen was born in Toronto, but grew up in Brockville, Ontario, where he began photographing trains in 1953. Through five years of university in Kingston, Don's interest in photographing Canadian railways continued to grow, and he found new horizons in 1963, when he began teaching secondary school in London, Ontario. In his own words, Don endeavours "to photograph Canadian trains in their setting because they complement the beauty of this country and are so much a part of the Canadian historical fabric."

William D. Miller
A third-generation railroad photographer, Bill Miller began taking pictures in 1980. Like his grandfather, the late William I. Miller, and his father, William E. Miller, Bill maintains a large collection of locomotive photos, backed up with extensive files and roster information. However, Bill spends more time out on the road, preserving railroad action on film, and has traveled extensively in pursuit of this goal.

Doug Phillips
Doug Phillips was born in Brandon, Manitoba and resided in Saskatchewan and British Columbia prior to moving to Calgary, Alberta, in 1960. An active photographer since 1965, Doug has amassed a personal collection of more than 40,000 railroad images. Noted for exceptional work, Doug's strongest interests are in western Canadian history and in the history of Canadian railways. For the better part of the past 20 years, he has worked for CP Rail and is currently supervisor of Maintenance of Way Training for CP's Heavy Haul System at Alyth Yard in Calgary.

George W. Roth
George Roth started watching trains as a young boy in Preston, Ontario, and began photographing images of the railway in his early teenage years. Working his way through university, George was employed in train service on the still-electrified Grand River Railway and Lake Erie & Northern out of his Preston hometown. Currently residing in Waterloo, George operates a graphic design consulting firm, and when time permits, he continues to watch and photograph railway operations in southwestern Ontario. Along with William Clack, George authored and helped illustrate a recent book on the CP Electric Lines. In addition, George's photographs, especially those of steam and electric operations in southern Ontario, have appeared in several books and other publications.

Larry Russell
A noted locomotive historian, Larry Russell has been an active railroad photographer since 1958. Larry has devoted considerable effort to the pursuit and preservation of information on the production, operation and history of North American-built steam, diesel and electric locomotives. In pursuit of this interest, Larry has travelled the continent extensively and has also journeyed overseas in search of locomotives exported by North American builders or those built to North American designs.

Robert J. Sandusky
One of the most respected Canadian railway photographers, Bob Sandusky began rail photography in 1946, thanks to a railfan mother and a high school beside the CNR Oakville Subdivision. A "student of anything on rails in any country," the Sandusky credit line has graced numerous books and other publications. Working in black & white, as well as colour, most of his photography has been concentrated in eastern Canada and the eastern U.S., with an emphasis on action and operational views. Bob is noted for having a sensitive eye and great ability to capture the human element of railroading. Photographically, his goals are "to create an historic record for the future and have fun doing it."

George Schaller
George Schaller has been interested in railways since his pre-school days. He received his first camera at age 10, and has been photographing trains ever since. George's grandmother lived just up the street from the CPR Galt, Ontario depot, and as a boy, he spent many hours there observing CPR steam, as well as the freight motors and passenger trains of the CP Electric Lines. Before graduating from University of Toronto as a pharmacist, George worked briefly at the Grand River Railway Preston Junction station pictured in this book.

Stan J. Smaill
Stan Smaill has expressed his lifelong interest in railroading, not only in photography, but in music, career choice and by taking an active role in the preservation and operation of first-generation Canadian diesels at the Canadian Railway Museum in Delson, Quebec. A train dispatcher for CP Rail in Montreal, Stan has been an active photographer for more than 30 years and began his railroading career as a high-school student working summers on CPR steel gangs in western Canada. Since then, he has also worked as a locomotive hostler for the Northern Alberta Railways and as an operator and train dispatcher for CP Rail on the Kootenay, Revelstoke, Smiths Falls and Quebec divisions.

Andrew J. Sutherland
Interested in railroading since his earliest memories, Andrew Sutherland began taking railroad photographs in 1970 with a fold-up Zeiss-Ikon that netted "biiiiig slides." Railroading interests led Andy to a 14-year career with CP Rail as an operator and train dispatcher at various points on the Kootenay, Toronto, Moose Jaw and Saskatoon divisions. Currently employed with GE Railcar Services, he was transferred from Calgary to Chicago in 1991. Andy continues to pursue photographic goals, which include preserving on film and sharing "memories of scenes which capture the links between railroading and the physical and economic geography of Canada," as well as images that stir "appreciation of the complexities and difficulties faced by railroaders in generating ton-miles . . . the human element behind the bald statistic."

John Sutherland
John Sutherland has been photographing Canadian railways for more than 25 years. While his earliest works were traditional "portrait-style" views of trains and locomotives, John quickly graduated to more artistic interpretations of Canadian railroading, paying special attention to details, history, and capturing Canadian railroads within the context of their environment. Having done time as an operator on CP Rail's Kootenay Division and spent university work terms on such projects as the 1972 relocation of the CP's Mountain Subdivision east of Rogers, British Columbia, John is currently employed by the CP Rail Engineering Department in Montreal.

James Walder
The son of a CPR engineer, the late James Walder grew up in Toronto and began taking railroad photographs during the 1940's. Working in black & white, as well as colour, Jim traveled extensively and his photographs have helped illustrate a number of books and other publications. The James Walder photographs reproduced in this volume were provided through the courtesy of his cousin, John Riddell.

John Welsh
A railroad photographer for more than 50 years, the late John Welsh was raised in Fergus, Ontario, and later resided in Toronto and Montreal. John made frequent business trips and always travelled by rail, using these opportunities to augment his extensive collection of black & white and colour photographs of U.S. and Canadian railways. Carleton Smith, curator of John Welsh's colour slides, provided the images used in this volume, while John's negatives are now included in the expansive Paterson-George Collection.

Richard Yaremko
Richard Yaremko resides in Calgary, Alberta, and has been active in railway photography since 1968. Although well travelled, he is at home roaming the Pacific Northwest, working to capture action on the BCR, CN's Edson and Albreda Subdivisions, as well as BN on Marias Pass and Montana Rail Link.

On November 19, 1966, radio days are still in the future as CP Extra 8600 West exits the siding at Puslinch, Ontario, following a meet with Windsor-Toronto passenger train #338. Having closed the siding switch, the tail-end brakeman swings back aboard steel van 437344 and waves a spirited hand signal to the distant head end... Highball!
John Freyseng